复兴混凝土

A Concrete Renaissance

[丹] BIG 建筑事务所 等 | 编
蒋丽 周荃 | 译

大连理工大学出版社

重塑生活方式：纽约的新型住宅模式

复兴混凝土

回收和再利用

C3 建筑立场系列丛书 No.79

C3 No.79 A Concrete Renaissance

Re-Living: New Residential Models for NYC

A Concrete Renaissance

Recycle and Reuse

纽约自由女神博物馆_EUS+ Architects + Folio: + Iowa State University

Jungwoo Ji（来自EUS+ Architects）、Bosuk Hur（来自Folio:）和Suk Lee（来自爱荷华州立大学）共同赢得了位于纽约的自由女神博物馆的设计竞赛。

自由女神博物馆应努力为人权和社会公正事业打造一个全新的世界理念和象征。它也应当成为能够引起公众对上述问题关注的标志性建筑，并为游客提供一次有关人权运动的各个方面和事件的解说式学习体验，成为一个反抗暴政的"虚拟中心"。

在韩国，民间烛光游行正在如火如荼地进行，以恢复民主价值、反对不平等的人权和社会不公现象。而项目团队设计自由女神博物馆的灵感就来自于此类民间运动的形势。

该博物馆坐落于纽约自由女神像所在的自由岛上，但它并不是纽约以及整个美国倡导人权和社会公正的唯一场所。这是一个能够实时反映当今世界人权状况和司法公正在人们日常生活中发挥的作用的建筑设施，也是这些活动的中心。

比如，当人们使用智能手机通过推特（一种社交网络）给这家博物馆发送信息，告知其所在城市的人权和社会公正方面的情形时，每个通过无线方式连接每一个地区的装置将接收到电子信号，并通过机械系统改变装置的角度，将装置的角度转向该区域。而装置的倾斜表明该区域的人权和社会公正的形势有所下滑。而更多的变化将使得自由女神博物馆的景观不断地发生改变。当然，建筑师希望看到这些博物馆装置一直是笔直朝向天空的，因为那就意味着整个世界的人权和社会公正情况良好。而鉴于这一点，自由女神博物馆被称为"社会公正媒体"。它是一个能够进行实时互动的新型社会媒体，而不仅仅是供游客参观展品的解说式博物馆。

相比自由女神像的巨大单体垂直纪念碑的传统形式，这个新型的"社会公正媒体"是综合运用水平状态、集合性、多样性和景观而创造出来的。设计师在这个地方提出了新型纪念碑的设计可能。博物馆背景中拥挤的纽约摩天大楼与该场所产生了联系。如果说自由女神像的火

设计过程
process

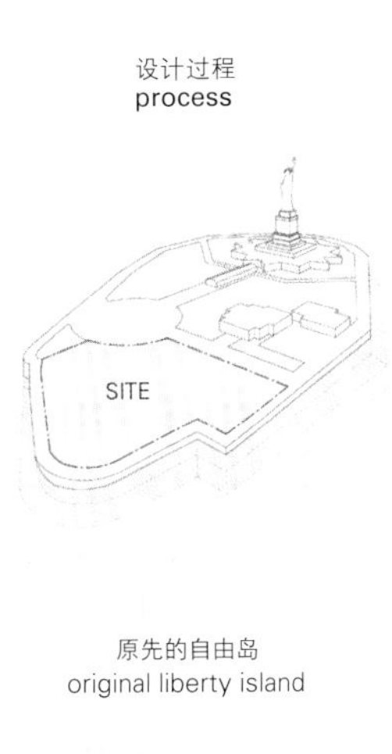

原先的自由岛
original liberty island

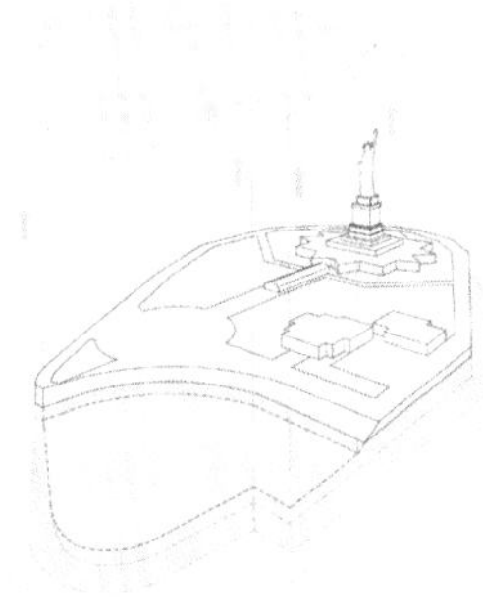

挪动的场地——露出水面
removed ground - exposed water

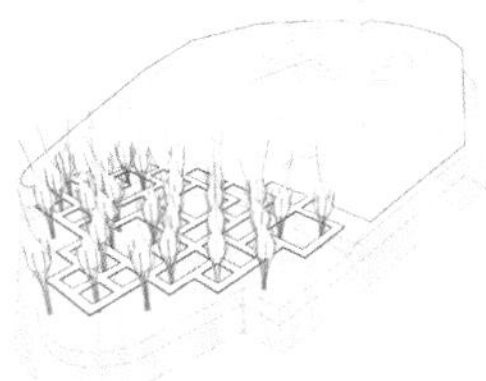

设计新增项目
proposing program addition

项目功能
program

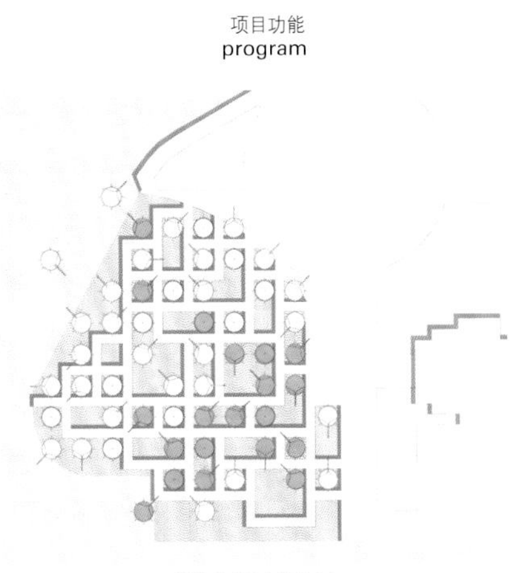

博物馆/多功能空间
museum / miscellaneous

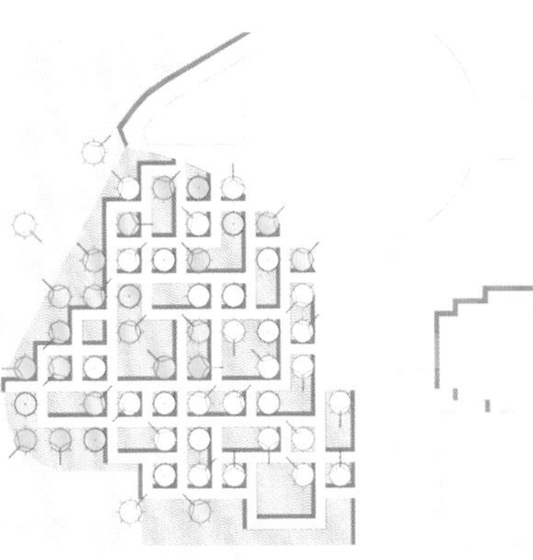

活动中心/多功能空间
action center / miscellaneous

社交与娱乐场所/多功能空间
social & leisure / miscellaneous

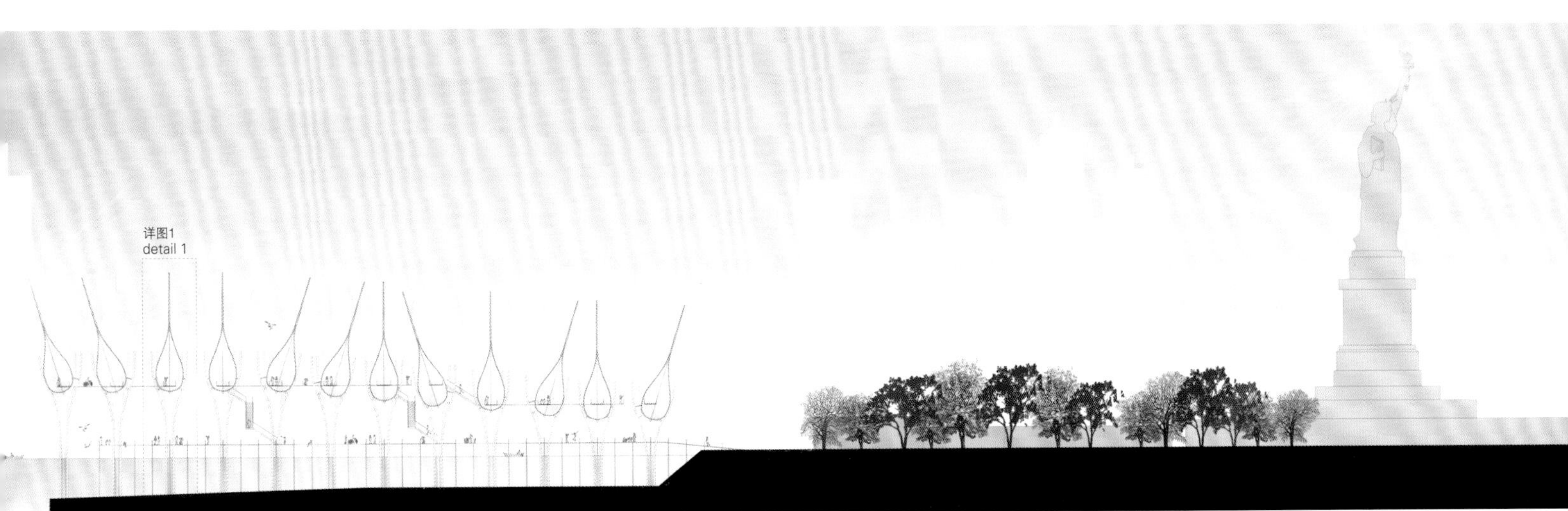

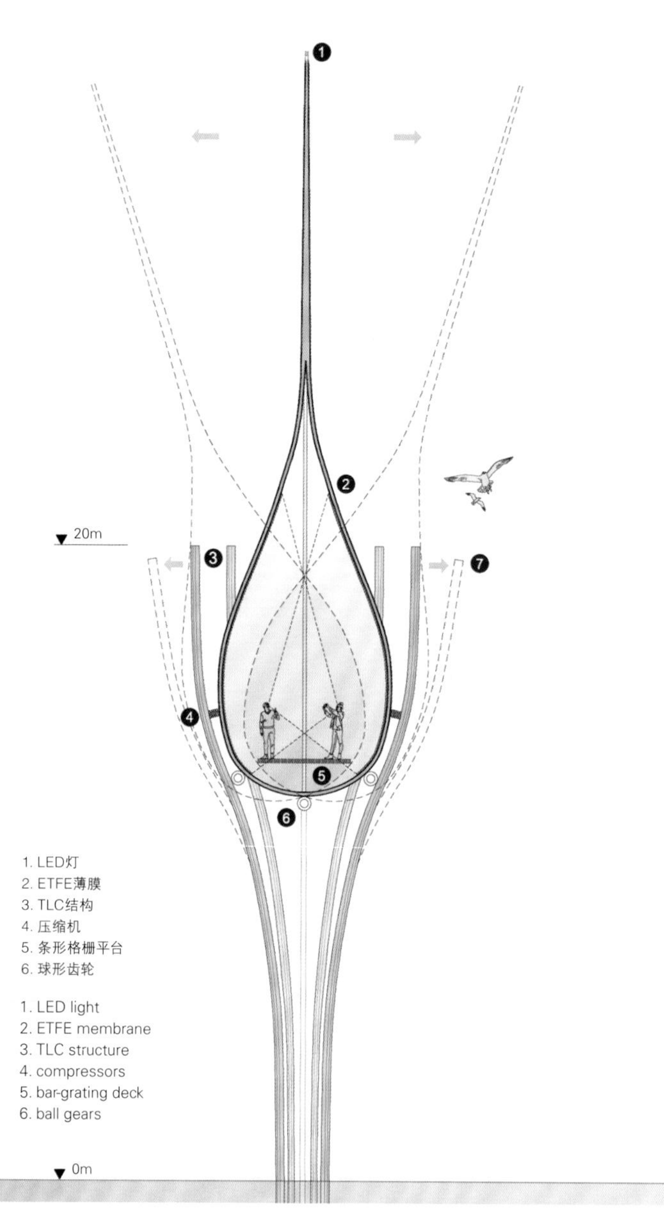

详图1 detail 1

炬代表了一种单一的理念，那么这里设计的数以百计的水珠就是测量不同价值观和不同地区的传感器。游客可以来到这里，在装置的下方，甚至可以在水面或是水面上空体验到这种测量结果，并且每一个装置都在投影面上实时显示世界每一个区域的人权和社会公正的情况。

为装置提供支承力的结构可能呈一小团火焰或是一小滴眼泪的形状，其设计灵感来源于在恐怖事件中倒塌的纽约市世贸中心双子塔。

Liberty Museum in New York

Jungwoo Ji (EUS+ Architects) and Bosuk Hur (Folio:) with Suk Lee (Iowa State University), have won a competition to design the Liberty Museum in New York.

The museum should strive to create a new-world idea and a symbol for the cause of civil rights and social justice. The museum should become an icon that would generate mass awareness on the aforementioned issues and provide an interpretive learning experience for the visitors on various aspects and events of the civil rights movement, becoming a "virtual epicenter" for the resistance against tyranny.

In Korea, the civil candlelight marches are in full swing to restore the value of democracy against unequal civil rights and social injustice. The project team has designed this museum inspired by the situation of such a civil movement.

This museum is located on Liberty Island with Statue of Liberty in New York but it is not the only place for citizenship and social justice in New York and the United States. It is an architectural device that reflects in real time how human rights situation and justice of the world are working in people's everyday life as well as the center for those activities.

When people send messages such as tweets to this museum about their city's human rights and social justice using their

smartphones, each unit connected wirelessly to each area receives the electronic signals and changes the angle of the unit by the mechanical system. The angle is towards that region. The tilt of the unit indicates that the decline in human rights and social justice in the region decreases. The more changes will make the whole landscape of the Liberty Museum become different continuously. Their wish is that these museum units are standing straight up all the way to the sky, which means the human rights and social justice of the whole world are in good situation. With this point, this Liberty Museum is called as "Social Justice Media". It is a social media of a new style as an object which mutually reacts in real time, not merely an explanatory museum to go and see the exhibits.
To the contrast of the Statue of Liberty which is a single, huge vertical monument of the former method, the new born "Social Justice Media" has been generated by horizontality, collectiveness, diversity, and landscape. The designers propose a possibility at this place as new typological monument. The congestion of the NYC skyscrapers in the background associated with this place. If the torch of the Statue of Liberty has represented a single idea, hundreds of water droplets here are measuring sensors of diverse values and various regions. Visitors can experience this down here, even on the water surface, even above, among others, and each unit shows the situation of human rights and justice in each region of the world in real time on the projection surface.
The structure that supports a unit which may be a small flame or a small teardrop is designed with inspiration of the pillar structure of the World Trade Center Twin Towers in New York which were collapsed by terror.

“穆赫兰的最后家园”项目的“矛盾住宅” _Hirsuta

建筑研究项目"arch out loud"与"穆赫兰的最后家园(LHOM)"项目合作，设计一处位于好莱坞标志下的未来居所。建筑师 Hirsuta 从 500 名设计师中脱颖而出，成为此次好莱坞建筑设计竞赛的获胜者。

因场地的显著位置，项目受到了各界的广泛关注。LHOM 项目旨在推动一项积极的任务，并成为未来居所建造和居住的范例。随着科技持续影响人们的日常生活，社会习惯和居住形式也将发生变化。

这栋房屋努力推进了实验性住宅的外围护结构设计。有一个球体低悬在单柱支撑的地板上，这种形式的构思源自于已建成的、带有更为完美的几何结构的圆形房屋。它就像理查德·福斯特设计的"圆屋酒店(1968 年)"一样旋转，但速度更慢，可能要一年或更长的时间。以这种方式，房子的许多侧面不断地重新组合，产生了新的视觉轮廓和立面，变成了一个不断变化的矛盾体。于是，试验性住宅不可避免的形象特征随着建筑的视觉效果和对建筑的解读方式的变化而受到了挑战。

围绕密集的固定核心区域，房屋的外部仿佛巨大天体厚厚的大气层一样在旋转。随着时间的推移，经过几个月或几个季度，旋转的空间和那些静止的空间都相应发生了变化。而通过这种方式，房屋内的生活方式也发生了变化。

入口层的平面既采用离心(楼板的外部区域)也采用向心(核心区)的布局方式，不同于任何一种单一的居住模式。房屋的底层主要用于容纳建筑的机械设备。公共事业设备如水、垃圾和燃气通过固定的核心区域进行运作，而电力实现了自给自足，单纯依赖于建筑的太阳光电外表皮产生的电能。建筑二层的线条更加鲜明，而三层完全是一个宽阔的密闭房间。在球状建筑的屋顶，这个几何形体创造了房屋周围和空中美景的集合。

建筑的大部分外表都由灵活的预制构件——光电薄膜包裹。光电薄膜本身就是建筑的外围护结构，而非附着于建筑的外围护结构之上，

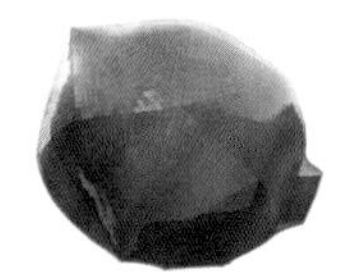
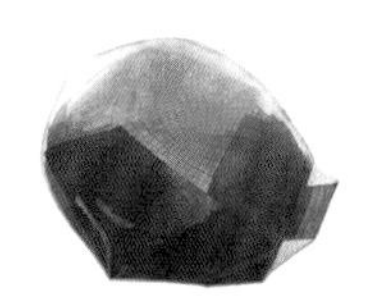

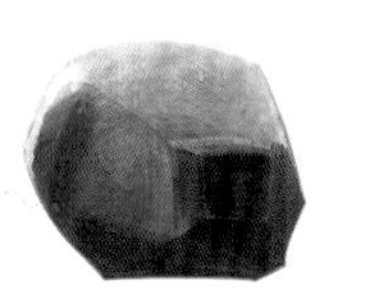
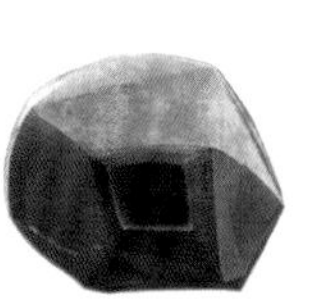

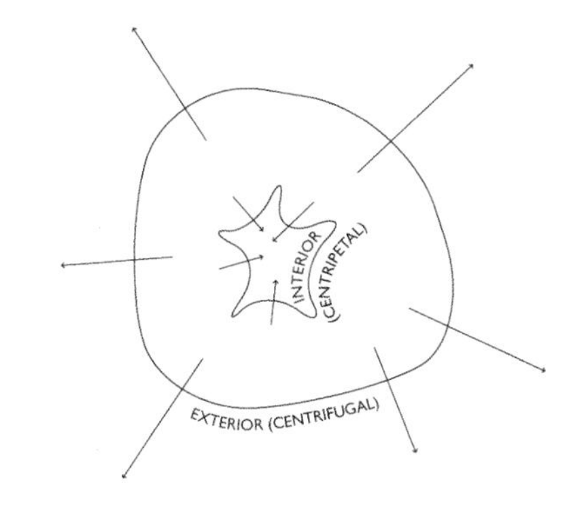

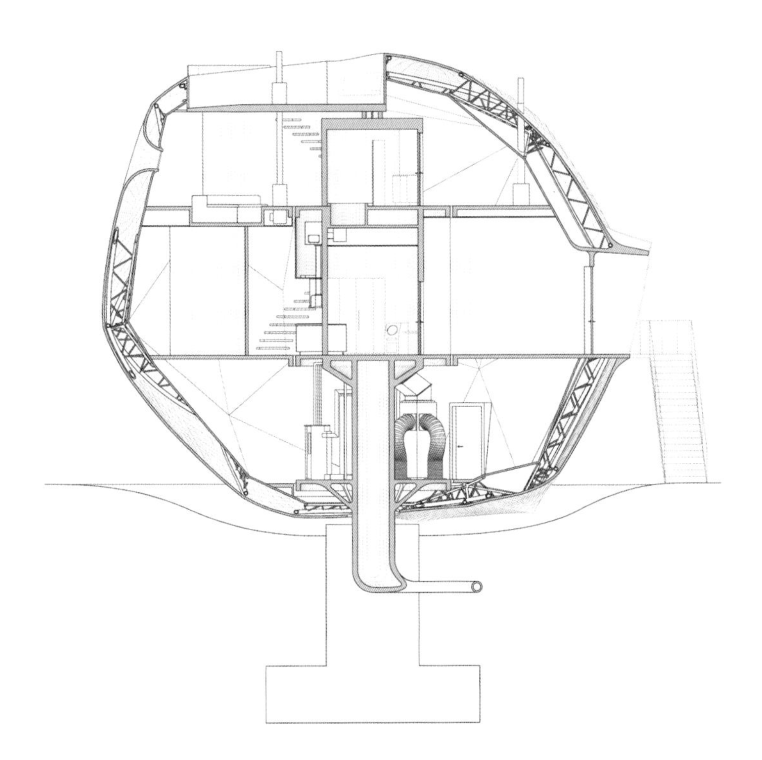

它将整个建筑包覆，不但实现了最大程度的日光照射强度，也使能源系统与建筑相结合。

该项目的房屋设计彰显了好莱坞山体的原野气息，但并未喧宾夺主。并不是所有的景观都能这样被归类：在一般的天气状况下，它并不是那么漂亮或别致，也不那么庄严，但是当它和山上的标志牌、岩石、灌木丛混合在一起时又显得那么浑然天成。

The Last House on Mulholland "Ambivalent House"

Architectural research initiative "arch out loud" partnered with "Last House on Mulholland(LHOM)" to design the house of the future, directly below the Hollywood Sign. The architect Hirsuta has been selected from 500 designers as the winner of the Hollywood architecture competition.

The location on such a prominent site enables the project to gain widespread attention. The LHOM project seeks to promote a positive mission and serve as an example for how future homes can be built and inhabited. As technology continues to impact daily life, social customs and living patterns will evolve along.

The house pushes hard on the envelope of experimental residential design. With a spheroid floating low to the ground on a single column, the form is an exact offspring of a more

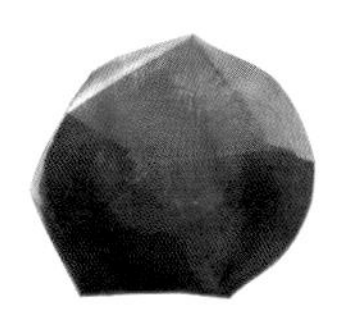
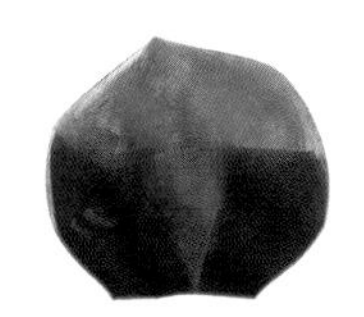
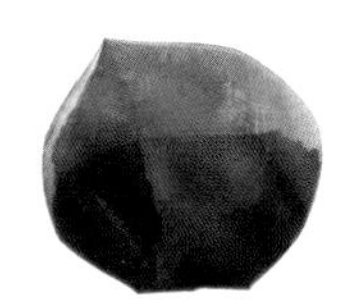

geometrically perfect round houses that are already achieved. It rotates, like the Richard Foster's "Roundhouse (1968)", but much more slowly, perhaps over the course of a year or more. In this way, the house's many faces continually recombine to produce new visual profiles and elevations, becoming an ever-changing, ambivalent object. The iconicity inevitable of an experimental house is challenged, then, by never being viewed or read the same way twice.

Around a dense, fixed core, the outer body of the house rotates as the thick atmosphere of a large celestial body. Over time, through months and seasons, spaces in rotation and those stationary change in relation. Through this, the lifestyle within differs as well.

The layout of the entry floor is organized both centrifugally (outer region of the floor plate) and centripetally (core), ambivalent to any single mode of occupation. The base of the house is dedicated primarily for the mechanics of the building. Utilities such as water, waste, and gas run through the fixed core, while electricity may be off-grid and rely solely on power generated by the building's photovoltaic skin. Stronger lines define the first floor, while a singular enclosed room opens up widely on the second floor. At the uppermost floor of the spheroid, the structural geometry creates constellations across the surrounding and overhead views.

The majority of the external building skin is cladded in photovoltaic film, flexible and panelized. Rather than being attached to the envelope, it itself becomes the envelope, wrapping the entire form to maximize solar exposure and integrate the energy system into the building.

The design embraces the wilderness of the Hollywood Hills and makes very little attempt to tame it. Very few landscapes can be categorized as such: not quite beautiful or picturesque, and not so sublime either, except under unusual weather. There is a certain brazenness as signposts and infrastructure mixing with rocks and shrubs so that each somehow becomes equal to the other.

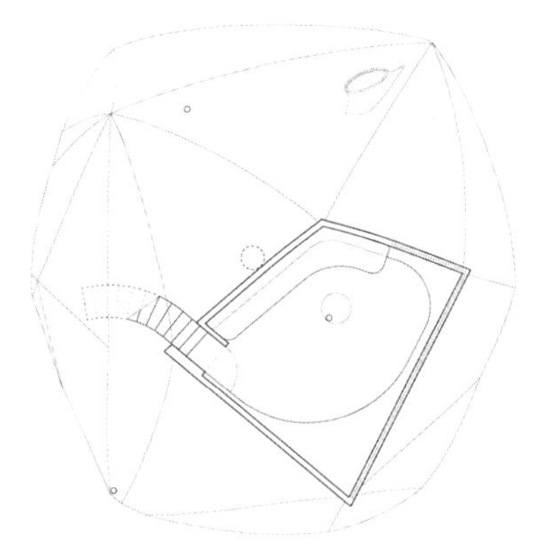

屋顶 roof

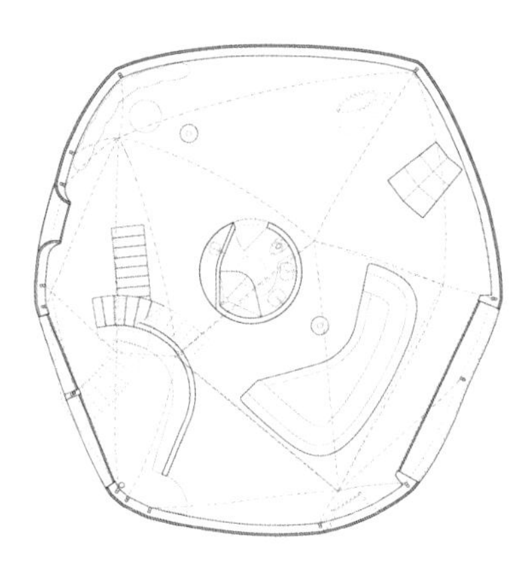

三层 second floor

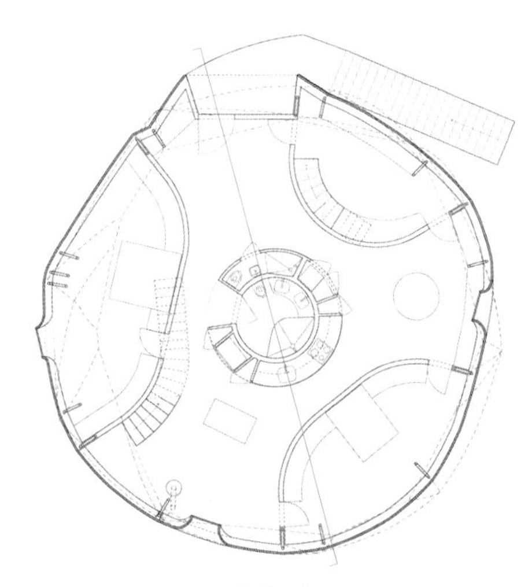

二层 first floor

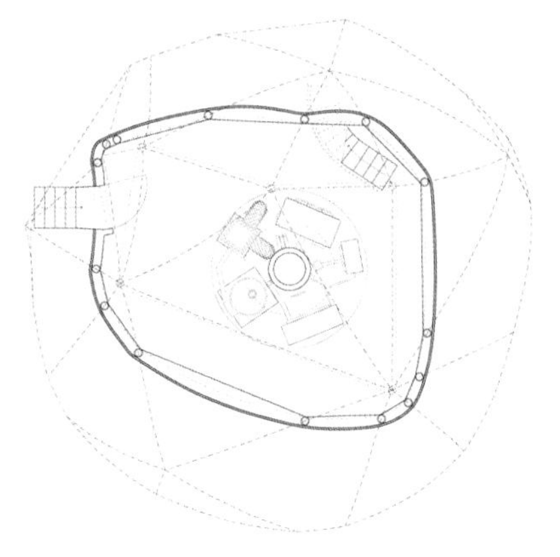

一层 ground floor

朝韩非军事区的地下公共浴室

人们将朝韩边境非军事区(DMZ)描述为"地球上非常可怕的地方"。这个实际存在的分界线的产生原因在于20世纪50年代朝鲜和韩国之间的矛盾，它是目前世界上戒备十分森严的边境线。尽管由于禁战，该边境地区相对平静，但因为它将朝鲜半岛一分为二，我们仍可以在呈对抗关系的两国之间感知其根深蒂固的紧张关系。由于边境地处偏远地区，因此造就了一个大型自然保护区，成为不少野生濒危物种的理想避难所。

建筑研究项目"arch out loud"就这一独特的区域举办了一次国际概念设计竞赛。主办方要求设计师们在朝韩边境非军事区内探求设计地下公共浴室的可能性，而且能够呼应周边的地理和政治条件。这种新型非军事型建筑能够占据该边境地区并开始缓解当前存在的紧张形势。旅游业能在联络边境关系方面发挥作用，但引出了一个问题：边境附近的建筑设计如何实现自身的定位？

设计竞赛的拟建场地位于第三条隧道的西侧，这条隧道总是接待大量的游客。此外，场地离开城工业园区（南北合作经济运营）的东南部不远。因此，运营期间工业园区的工人和来DMZ附近旅游的游客将会是浴室未来的顾客群。

在来自世界各国的将近300个设计提案和900多名参赛者中，Studio M.R.D.O. & Studio LaM（来自美国）设计的"穿越三八线"最终成为获胜项目，评委会同时还评出了5个获得亚军和10个获得荣誉奖的提案。竞赛评委会的成员包括Stan Allen、Moon Hoon、Jing Liu、Lola Sheppard、Minsuk Cho、Kristy Balliet、Anna Neimark、Seunghyun Kang、Nicholas Bonner、Yehre Suh和Matias Del Campo。其评选标准主要基于作品如何通过空间功能、场景和叙述手法来表达边境区域的冲突。面对这个敏感地区的疑难问题和事件，建筑师所提交的项目向评委会展示了诸多充满诗意而又庄严的设计方法。

Korean Demilitarized Zone Underground Bathhouse

Demilitarized Zone(DMZ) is an area that has been described to be the "scariest place on earth". This de facto barrier, which manifested in the 1950's as the result of conflicts between the Democratic People's Republic of Korea and the Republic of Korea, stands today as the most heavily fortified border in the world. Although the site is relatively calm (since no combat is permitted), deep tensions can be sensed between the rivaling countries, as it splits the Korean Peninsula into two sides. Due to the border's remoteness, a great natural reserve is created where several endangered species of wildlife take ideal refuge. Architectural research initiative "arch out loud" has opened an international open-ideas competition for this unique area. The host challenges designers to explore the possibility of creating an underground bathhouse within the DMZ which responds to the surrounding geopolitical conditions. New forms of non-military architecture could occupy this border zone and begin to ease the existing tension. Tourism can play a role in opening relations across a border that begs the question: How does architecture position itself in the middle of this? The proposed site area for the competition is located just west of the third tunnel which receives a high volume of tourist traffic. In addition, the site is located not too far southeast of the Kaesong industrial park, a collaborative economic operation between both the North and South. Thus, the bathhouse will be used both by workers of the industrial park when it resumes operations and visitors of nearby DMZ tours. Out of nearly 300 proposals and over 900 participants from all over the world, "Crossing Parallel(s)" by Studio M.R.D.O.& Studio LaM (based USA) as the winner project, 5 runner-ups and 10 honorable mentions were selected. The jury consisted of Stan Allen, Moon Hoon, Jing Liu, Lola Sheppard, Minsuk Cho, Kristy Balliet, Anna Neimark, Seunghyun Kang, Nicholas Bonner, Yehre Suh, and Matias Del Campo. Criteria were mostly based on how conflicts were addressed through spatial programs, scenarios, and narratives. Submitted projects showed a variety of poetic and sublime approaches for confronting the difficult questions and issues of this sensitive zone.

朝韩边境非军事区 Korean demilitarized zone

柏林墙 Berlin wall

美国-墨西哥边境线 U.S. - Mexico border

都罗展望台 Dora observatory

共同警备区 joint security area

桑拿结束 finish sauna

曼哈顿地下城市，奥斯卡·纽曼 underground city beneath Manhattan, Oscar Newman

穿越三八线：作为隐喻式剧院的公共浴室 _ STUDIO M.R.D.O. & Studio LaM

三八线不是一条表面上细细的线，而是一种具有更深意义的存在：在朝鲜和韩国的徘徊于紧张和松弛之间矛盾的情绪积累之下，它已凝固。

在名为"隐喻式剧院"的设计提案中，来自两边的游客（演员/观众）行走于双螺旋坡道之上，重现了这种凝固的过程。虽然更多的是彼此远离，但仍然有合并和分离、跨越不可跨越的界线那一瞬间的体验。游客一旦抵达公共浴池，所有这些体验就都化作液体流入水中，游客所带来的情感碎片也浸入彼此的皮肤之中。

Crossing Parallel(s) : Bathhouse as a Metaphorical Theater

The 38th parallel is not a thin superficial line, rather a thickened situation: it has been solidified by accumulation of ambivalent emotions - tensions and relaxations - between North and South.

In the proposed bathhouse, represented as a "metaphorical theater", visitors (actors/audiences) coming from each side reproduce the process of such solidification while walking down the double helix ramp; experience of merging and diverging, moments of crossing uncrossable lines, while being more away from each other. Upon reaching the communal pool, all such experience is liquefied into water, and debris of emotions brought by visitors soaks into each other's skin.

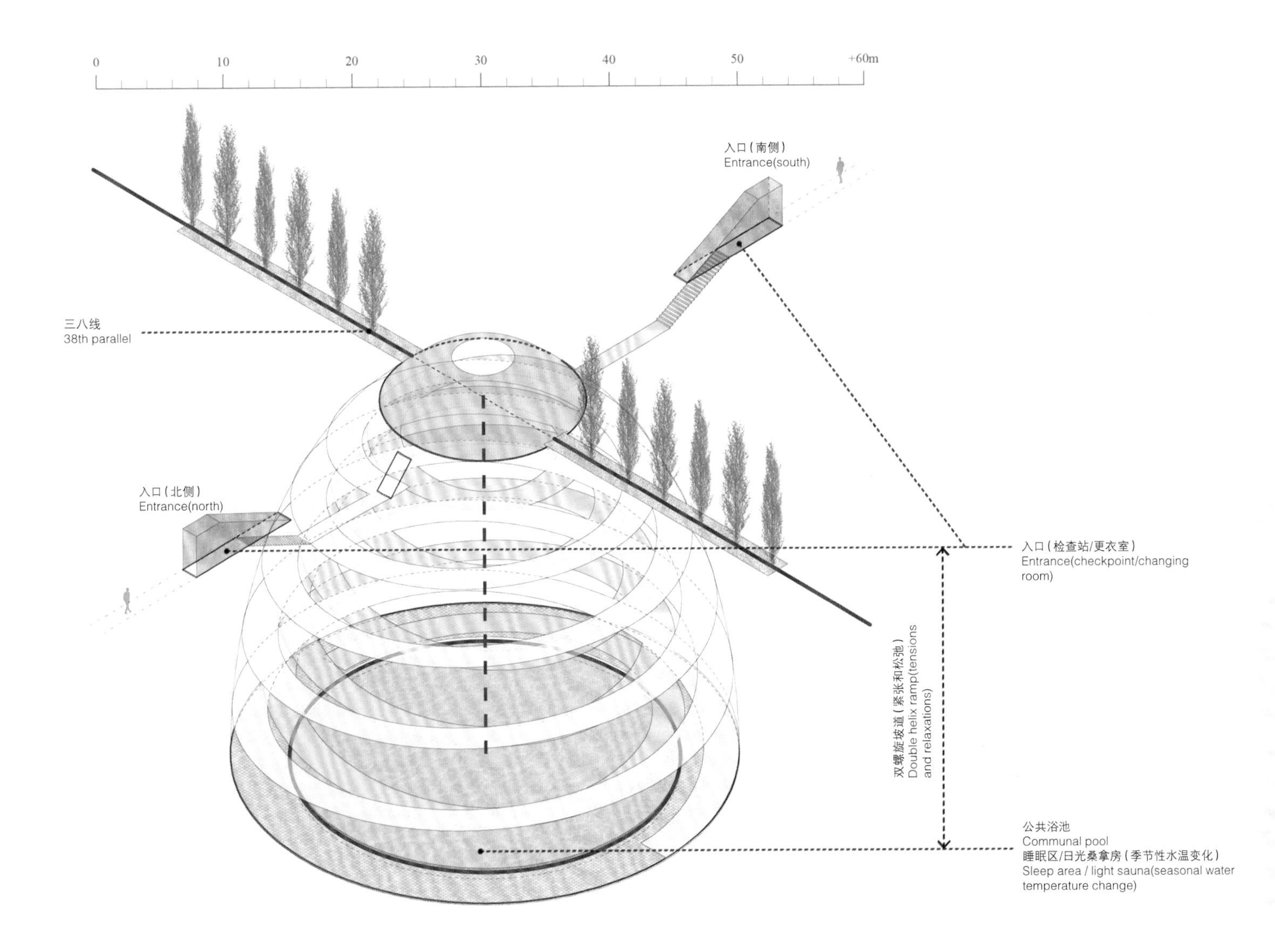
0
10
20
30
40
50
+60m
入口(南侧)
Entrance(south)
三八线
38th parallel
入口(北侧)
Entrance(north)
入口(检查站/更衣室)
Entrance(checkpoint/changing room)
双螺旋坡道(紧张和松弛)
Double helix ramp(tensions and relaxations)
公共浴池
Communal pool
睡眠区/日光桑拿房(季节性水温变化)
Sleep area / light sauna(seasonal water temperature change)

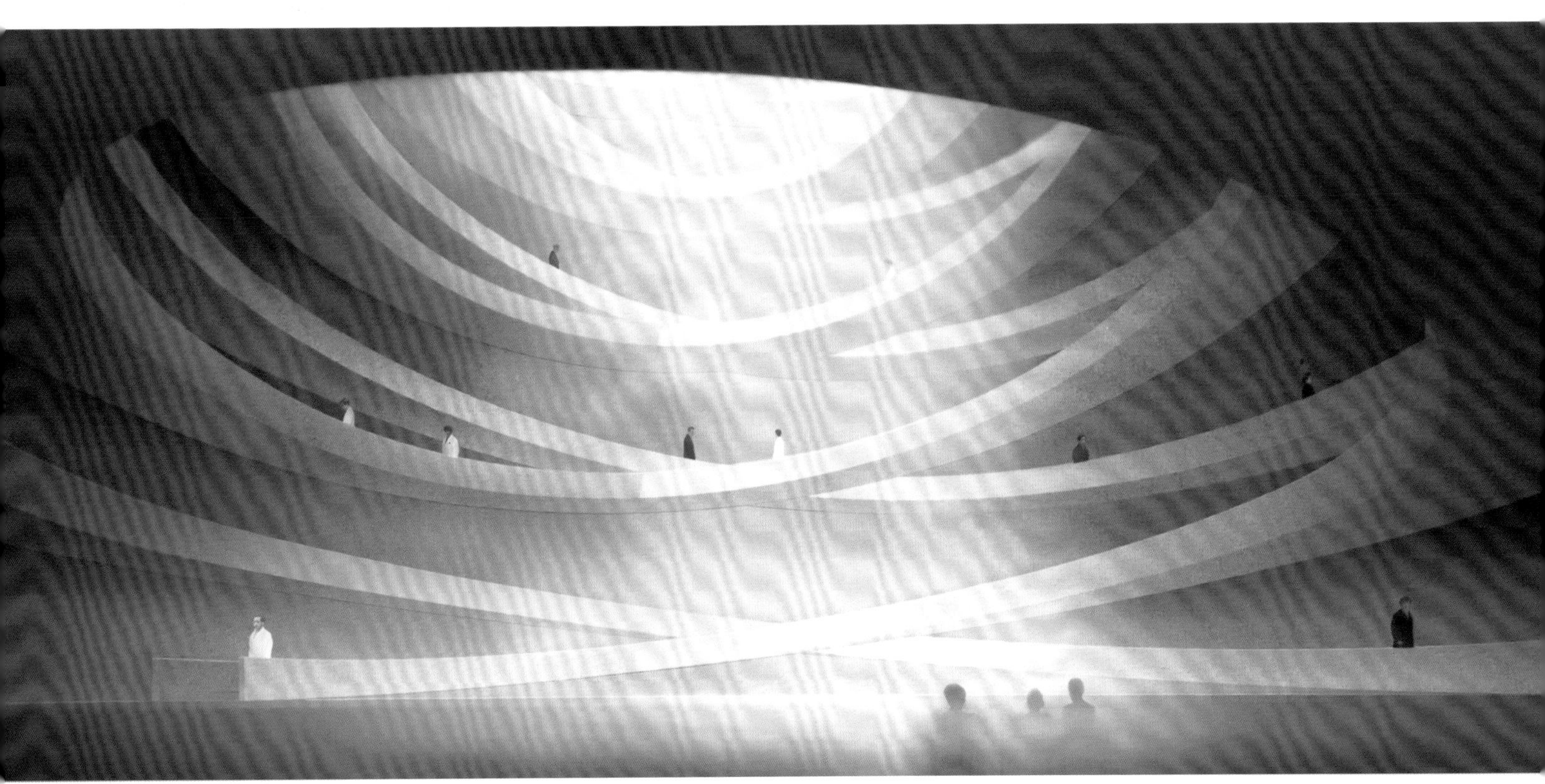

高远的天际 _ Vuk Filipic + Anna Murynka

朝韩边境非军事区（DMZ）极端的紧张气氛矛盾地加强了一种紧张之下的平静。与这个地区的设立目的相似，该项目通过一种反思的过程来消除双方之间的矛盾。来到公共浴室的客人平等分享孤独的感觉和内在的反思，不管他们在外面的世界里是什么出身或社会地位。悬吊的玻璃容器使客人悬浮于巨大的地下穹顶中，在这里他们可以享受蒸汽、热水浸泡和冷水冰浸的过程。这是一个呈现人们同时被隔离的巨大场景，反映了浴室之外的人类状况。

This Lofty Sky

The extreme tension of the DMZ paradoxically reinforces an intense peace. With a similar intention, this project proposes to erase the interpersonal conflict through a process of introspection. Guests of the bathhouse share in the equality of solitude and inner reflection, regardless of their origins or state of existence in the external world. Suspended glass vessels lower individual bathers into a vast subterranean dome, treating them to a steam, hot soak, and paralyzing ice dip in the process. A mass spectacle of simultaneous personal isolation unfolds, mirroring the human condition beyond the bathhouse.

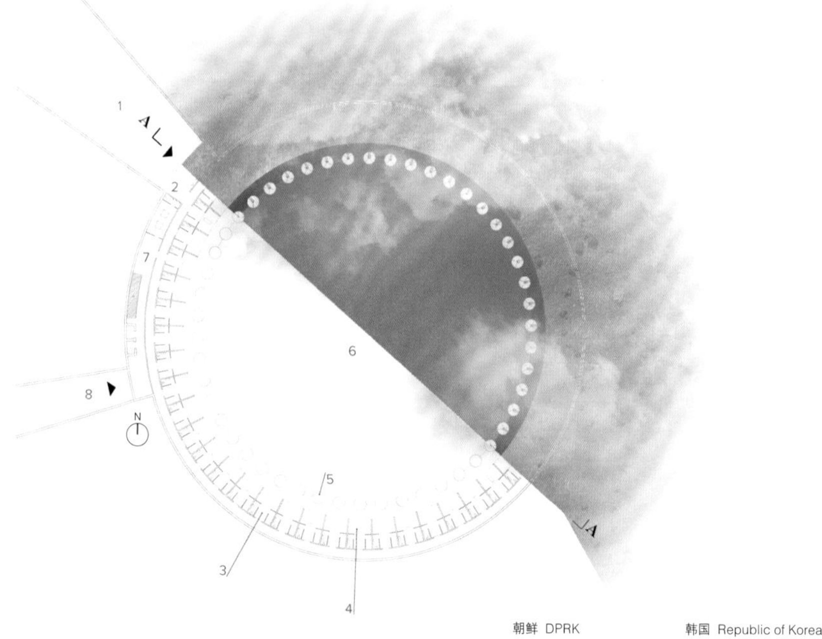

1 入口坡道
2 门厅
3 更衣室/淋浴房/卫生间
4 按摩房/睡眠区
5 下降的容器（独立热水池）
6 冰水池
7 员工区
8 装卸台

1 Entrance Ramp
2 Lobby
3 Changing/Shower/WC
4 Massage/Sleeping
5 Lowering Vessel (Self Contained Hot Pool)
6 Ice Pool
7 Staff Area
8 Loading Ramp

朝鲜 DPRK　　韩国 Republic of Korea

跨越 _ Xiaoyu Wang + Yutian Wang

我们的项目设计目的是探索边境如何作为关键的元素发挥作用，将两个彼此对立的地区融入到一个统一的整体中，而非将它们分离。通过引入一条在两国之间舞动的波状线条，一面不断交织的地下墙体连接一系列共享的开放浴池和坚实的单间，其圆形的几何结构承袭了传统韩国澡堂——韩式桑拿房的构造原理。通过将自然光和地形景观从地上引入到地下，建筑师创造了多种跨越边境线的间隙环境，以此鼓励来自朝鲜和韩国的人们展开实际的相互交流。

Cross

Our project aims to explore how the border functions as a key element that embraces two contradicted territories into one united entity rather than separates them. By introducing an undulated line that dramatically dances in-between the two countries, a continuously weaving underground wall ties a series of collectively shared open pools and solid individual rooms with round shape geometry which is inherited from the mechanism of traditional Korean bathhouse typology Jimjilbang. By inviting natural sunlight and topographical landscape from above-ground to underground, this form creates multiple crossing-border interstitial conditions that mutually encourage people from both DPRK and Republic of Korea to interact physically.

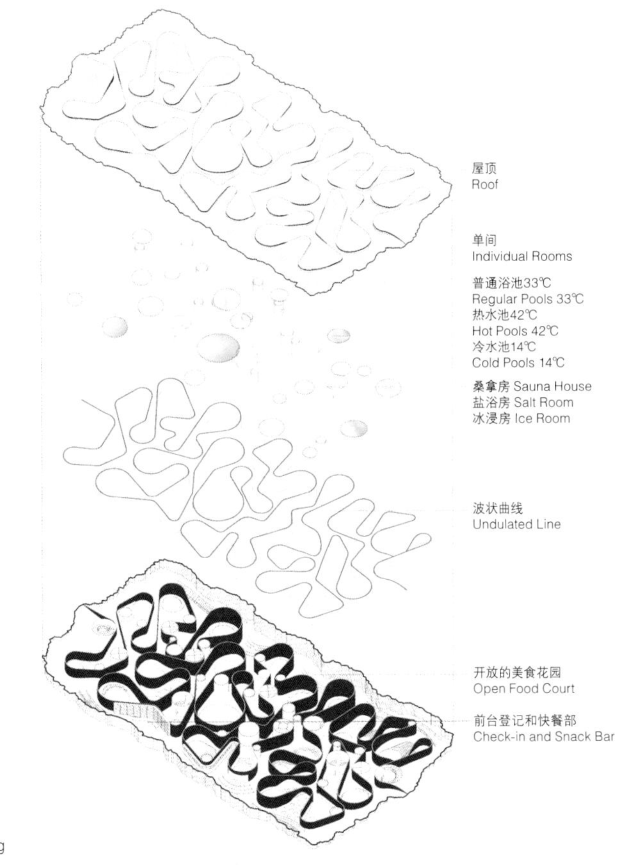

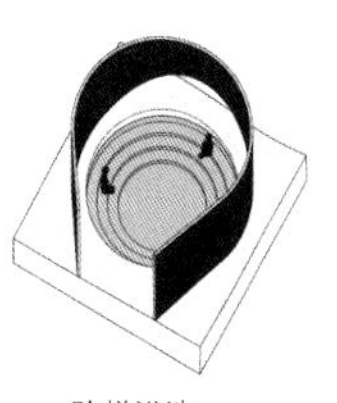

阶梯浴池
Stepped pool

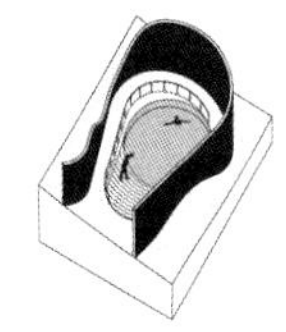

坡道浴池
Sloped pool

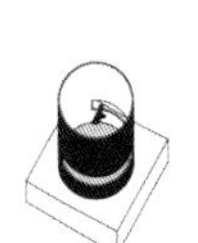

单人浴池
Individual pool

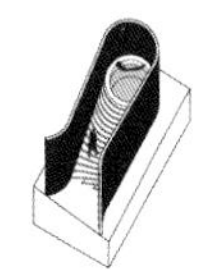

升高浴池
Elevated pool

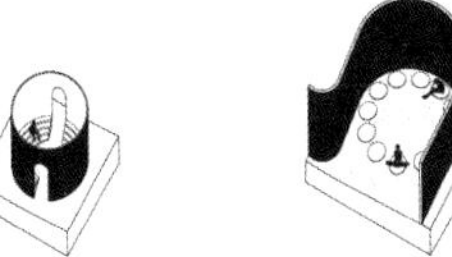

桑拿房
Sauna room

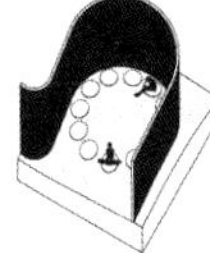

休息室
Resting room

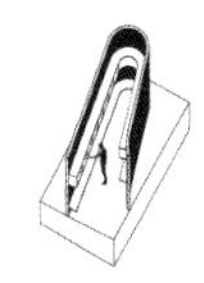

更衣室
Changing room

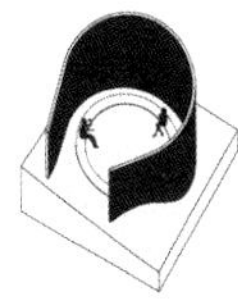

有连续座位的休息室
Lounge with continuous seating

重塑生活方式
纽约的新型住宅模式

Re-Livi
New Residential

曼哈顿岛的空间有限，却无时无刻不在进行施工。城市密度一直是许多概念方案的主题，但新的建筑方法往往很难有机会实施。相反，城市简单地持续向垂直方向上升起，而且像Rafael Viñoly设计的公园大道423号这样的新住宅建筑为城市带来了一种泛泛的——尽管有时美得平庸——居住体验。而本文介绍的两个新项目证明了，目前也许还不是选择另一种生活方式的最终时刻。

曼哈顿的特里贝克区附近的行人活动量较大；铺设鹅卵石的街道使车辆不得不减速慢行，而无处不在的人行道边门前露台也鼓励人们在此休息并且享受该地区的活力。目前，特里贝克区的街道已被纽约市地标保护委员会指定为历史街区，街道两旁壮观的铸铁立面林立，因拱门和装饰品不同，所以每个立面都很独特。在这个曾经的工业街区，先前的六层高的工厂和仓库被改造成豪华的顶楼寓所，而赫尔佐格&德梅隆设计的伦纳德大街56号公寓好似一个孤独的巨人巍然挺立。凭借较小的占地面积和参天的高度，该建筑已成为特里贝克区天际线的一大亮点，作为市中心高楼林立的世界贸易中心和位于中城区密集的办公楼之间的桥梁，高高耸立。

Manhattan is an island of finite space and endless construction. Its density has been the subject of many conceptual proposals, but it is not often that new approaches have an opportunity to be built. Instead the city has simply continued to rise vertically and new residential buildings like Rafael Viñoly's 432 Park Avenue have brought to the city a generic – albeit at times beautifully banal – residential experience. But two new projects test whether it might not finally be time for a different model of living.

The city's Tribeca neighborhood is rich with pedestrian activity; its cobblestone streets force vehicles to slow down and the ubiquity of sidewalk stoops encourages people to take a rest and enjoy the area's dynamics. Tribeca's streets, now designated a Historic District by New York City's Landmark Preservation Committee, are lined with spectacular cast-iron facades, each unique in the rhythm of its arches and adornments. In this once industrial

ng Models for NYC

西57号街超高公寓_VIA 57 West/BIG

伦纳德大街56号公寓_56 Leonard Street/Herzog & de Meuron

重塑生活方式：纽约的新型住宅模式

Re-Living: New Residential Models for NYC/Julia van den Hout

但与该城市许多新建的豪华大楼不同，凭借光滑连续的立面和偶尔带有顶楼露台的楼层设计，赫尔佐格&德梅隆的伦纳德大街56号公寓为人们提供了一种新选择：我们能否居住于高层建筑，将整个曼哈顿岛的景色尽收眼底（可能还会欣赏到纽约其他几个区的景色），适当地融入个性化特征，并且减轻刻板而毫无特征的建筑物中令人无法辨认的公寓单元的重复性。最终形成了一座由相互错落堆叠的玻璃盒子构成的、如实体动画一般的超豪华建筑，建筑的外围护结构由交替伸出立面的阳台和露台构成。

在中城区，毗邻住宅和商业大厦楼群中心的地方，丹麦BIG建筑事务所采取了相似的设计方法。而该公司的首个纽约项目——西57号街超高公寓（VIA 57 West），并未承受特里贝克区的历史因素造成的负担，而是受到相反情况的挑战。VIA项目位于曼哈顿的西边，临近曼哈顿西区高速公路，其设计任务在于为这座城市最快速的一条交通线路带来居住的平静。BIG试图在该项目中加入一些行之有效的丹麦式生活方式；VIA项目占据了一整条城市街区，勾勒出街区的周界，并空出内部空间作为公共庭院。这座城市绿洲从街面升起，首层是商用和文化空间，这一空间的作用在于将公寓与高速公路的喧嚣隔离开来，同时还可以一览哈德逊河的景

neighborhood, where six-story former factories and warehouses have been transformed into luxury lofts, Herzog & de Meuron's tower at 56 Leonard rises like a lone giant. With its slim footprint and immense height, it has become the single high point on the Tribeca skyline, standing tall as a bridge between the verticality of the World Trade Center downtown and the dense cluster of office towers in Midtown.

But unlike many of the new luxury towers rising in this city – with sleek continuous facades and an occasional penthouse terrace level – with 56 Leonard Herzog & de Meuron offers up a new alternative: what if we can have our serving of highrise living with its sweeping views over the entire island of Manhattan (and presumably also well into the boroughs), with a healthy dose of individuality, a relief from the repetition of unidentifiable units that lie behind stark, anonymous exteriors. The result is an ultra-luxury pixilation of stacked glass boxes, with an envelope

世界贸易中心
World Trade Center

公园大道423号，Rafael Viñoly
432 Park Avenue, Rafael Viñoly

色。与BIG位于哥本哈根新城区欧瑞斯塔的开发项目一样，每个住宅单元都采用向外倾斜的设计，以优化日光照射水平。

而VIA项目同时也证明了纽约人确实喜欢登高远眺和放松心情。项目采用内廊式设计，只有中庭一侧的公寓可以享受它带来的宁静。建筑外侧的单元面向狭窄的小巷，小巷里总有车流在焦躁地等着拐入高速公路。令人欣慰的是，VIA项目向住户提供了众多能够提高住房销售量的便民设施（从游泳池、篮球场到电影放映室和棋牌室）。这些日光充足的公共空间提供了大多纽约居民无法获得的资源：社交互动，以此将住户对公寓的熟悉和舒适性延伸至小空间之外。

伦纳德大街56号公寓的特征还包括不少令人印象深刻的便民设施空间——大楼的九层和十层容纳了图书馆、瑜伽馆、游泳池、桑拿房和剧院等舒适环境——朴实的露石混凝土室内环境和如雕塑一般的壮观楼梯使得这些区域不太像一个极度活跃的技术服务处，而更像是一个高端的度假胜地。这里的居民不太可能会像VIA项目的居民那样获得相同程度的邻里友好接触。但以数以百万的销售价格来看，伦纳德大街56号公寓的业主不太可能会在意这一点。赫尔佐格&德梅隆为每一套公寓都设计了不同的楼层平面图，为每一个住户都提供了一种独特的个人居住空间体验。

defined by the playful push-and-pull of balconies and terraces.

In Midtown, adjacent to the heart of residential and commercial towers, Bjarke Ingels Group has taken a similar approach. While the firm's first New York project, VIA 57 West, is not burdened by the historic parameters of Tribeca's pedestrian-heavy context, it is challenged with an opposite condition. Located on the western edge of Manhattan, abutting the West Side Highway, VIA is tasked with bringing residential calm to one of the city's most fast-moving traffic strips. BIG has tried to insert here a bit of its tried-and-proven Danish approach to living; VIA takes up a full city block, outlining the perimeter and freeing up the interior for a communal courtyard. This urban oasis is lifted from the street by a commercial and cultural ground level, separating it from the highway, and giving it uninterrupted views over the Hudson River. Like in BIG's developments in Ørestad, individual units are articulated by an outwards tilt, optimizing sun exposure.

However, VIA is simultaneously evidence of the reason why New Yorkers prefer to travel vertically for views and respite. With a double loaded corridor plan, only the apartments lining the courtyard benefit from its tranquility. The units on the outside of the building face the narrow side streets, where the traffic impatiently awaits its turn to merge onto the highway. As a consolation, VIA does provide its residents with a number of highly sellable amenities – from a swimming pool and a basketball court to movie screening and poker rooms. These light-filled communal spaces offer what most residences in New York cannot achieve: social interactions that extend the familiarity and comfort of your apartment outside of the confines of your small space.

While 56 Leonard also features impressive amenity spaces – the tower's ninth and tenth floors house luxuries like a

©Nic Lehoux

西57号街超高公寓
VIA 57 West

归根结底，便民设施只是附加物。它们不与城市产生相互作用，当我们路过时，它们并不形成我们对建筑的直接体验。因为面向河流，路过的车辆势必会看到VIA项目，这对于大多数纽约人来说是一个与众不同的视角。它表现为一个供人们眺望城市风景并且远离城市喧嚣的观景平台，而非城内的供人们观赏的对象。而凭借建筑高度和整体设计风格，伦纳德大街56号公寓两者兼具。人们拭目以待它将如何与地面连接，如何将融入其东北角的阿尼什·卡普尔雕塑包含在内，其东北角局部收拢于混凝土板的下方并向外鼓胀、跨越人行道。但因为明显高于城市中的多数建筑，它已然成为下曼哈顿区的一座醒目的标志性建筑。

这两个项目基本上是对于纽约这种大城市的生活方式的两种截然不同的诠释；项目由两家不同的欧洲设计公司为两种不同阶层的居民设计。VIA项目像是在试图实现丹麦风格的民主化空间，而伦纳德大街56号公寓则在宣扬奢华而与世隔绝的生活方式。VIA项目散发着青春和热情的光辉，而伦纳德大街56号公寓则展现了沉思和成熟老练的一面。但是，这两个项目的共同点在于尺度的变化——形成单元之间的接合，并且侧重于我们作为个体如何与内部个人居住空间以及家门外的整个城市发生相互作用。

library, a yoga studio, a swimming pool, a sauna, and a theater – the stark exposed concrete interiors and imposing sculptural stairs make these areas feel less like a hyperactive tech office but more like a high-end resort. Residents aren't likely to find the same friendly contact between neighbors as in VIA. But at sales prices in the many millions, buyers at 56 Leonard are not likely to care. With apartments that each have a different floor plan, Herzog & de Meuron offers every resident a unique experience of individual residential space.

And in the end, amenities are just extras. They don't interact with the city; they aren't what we experience of these buildings as we pass by. Oriented towards the water, VIA is meant to be seen by car, an unusual viewpoint for most New Yorkers. It presents itself as a platform from which to look out and get away from the city, rather than as an object within the city at which to look. With its height and design in the round, 56 Leonard offers both. It remains to be seen how 56 Leonard meets the ground, how it will embrace the Anish Kapoor sculpture that will be integrated into its north-east corner, partially tucked underneath the concrete slab and ballooning out across the sidewalk. But it has already become a visible landmark in Lower Manhattan, visible above much of the city's fabric.

Ultimately, these are two very different interpretations of living in a city like New York; they are designed by two European firms and for two different demographics of residents. What in VIA feels like an effort in Danish democratization of space, and in 56 Leonard becomes an advertisement for lavish isolation. Where VIA exudes young and enthusiastic brightness, 56 Leonard presents brooding and mature sophistication. But what these buildings have in common is a shift in scale – an articulation of the unit and a focus on how we, as individuals, interact with our personal residential space inside, and the city at large outside beyond the thresholds of our doors. Julia van den Hout

西 57 号街超高公寓

BIG

sanitation
EXIT
ALFALFA

西57号街超高公寓（VIA项目）是欧洲外围街区建筑与传统曼哈顿高层建筑的混合体，它同时具备两者的优势：既拥有庭院建筑的简洁与高效，又不乏摩天大楼的优良通风和广阔视野。建筑的东北角直冲云霄，最高处达137.2m，而另外三个角则顺势而下，保持在较低的高度之上，庭院的设计打开了面向哈德逊河的视野，也使西方的斜阳可以深入照射至建筑当中，并且优雅地保留了相邻的海伦娜大楼面向河流一侧的视野。建筑外形的变化主要由观者的视角来决定。从西侧高速公路看去，它就像一座金字塔；而从西58号街看去，该建筑则变成了一座引人注目的玻璃尖塔。建筑中庭的设计灵感来源于传统的哥本哈根城市绿洲，人们可以从街道看到中庭，而它起到的作用则是将附近哈德逊河公园的绿茵延伸至VIA项目中。建筑的坡度形成了南部低层建筑与场地北部和西部高层住宅建筑之间规模上的过渡。明显可见的倾斜屋顶由穿插露台的简单直纹曲面构成——每个露台都是独一无二的并且朝南。墙面的鱼骨图案同样体现在建筑的立面上。每一间公寓都设计有一扇飘窗，以扩大景观视野，而且都有阳台，鼓励居民与路人之间的互动。

VIA 57 West

VIA 57 West is a hybrid between the European perimeter block and a traditional Manhattan high-rise, combining the advantages of both: the compactness and efficiency of a

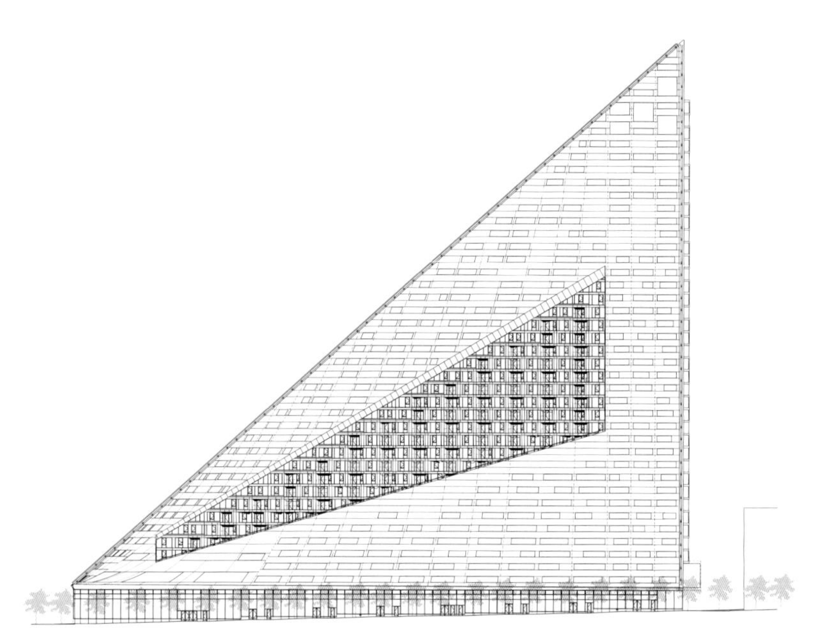

西南立面 south-west elevation

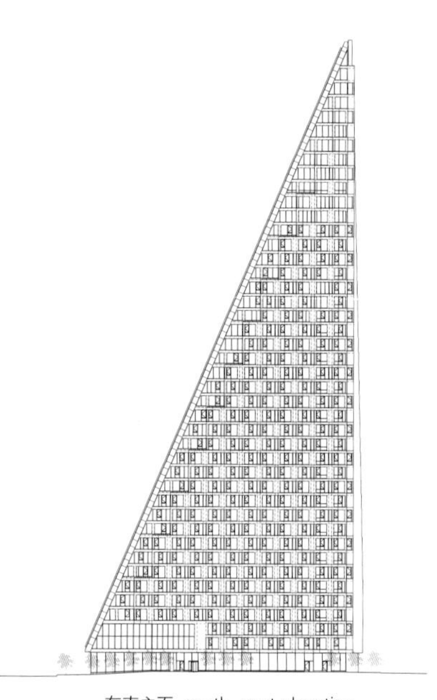

东南立面 south-east elevation

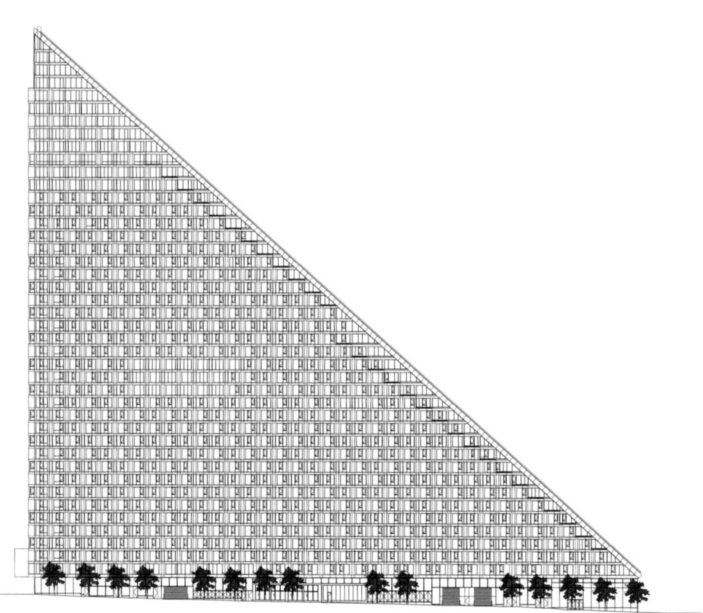

东北立面 north-east elevation

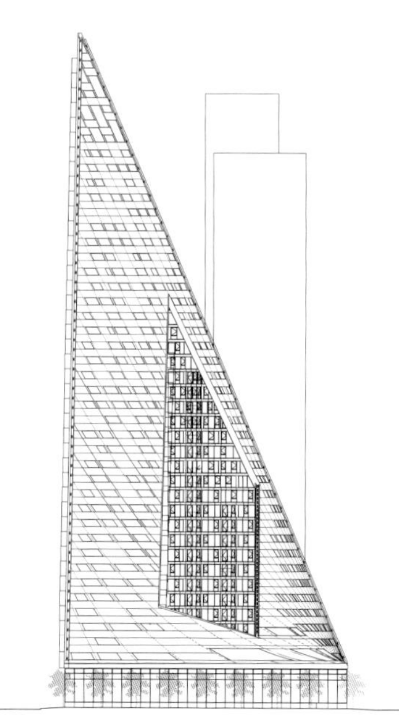

西北立面 north-west elevation

Coffee
Place
Lunch Menu

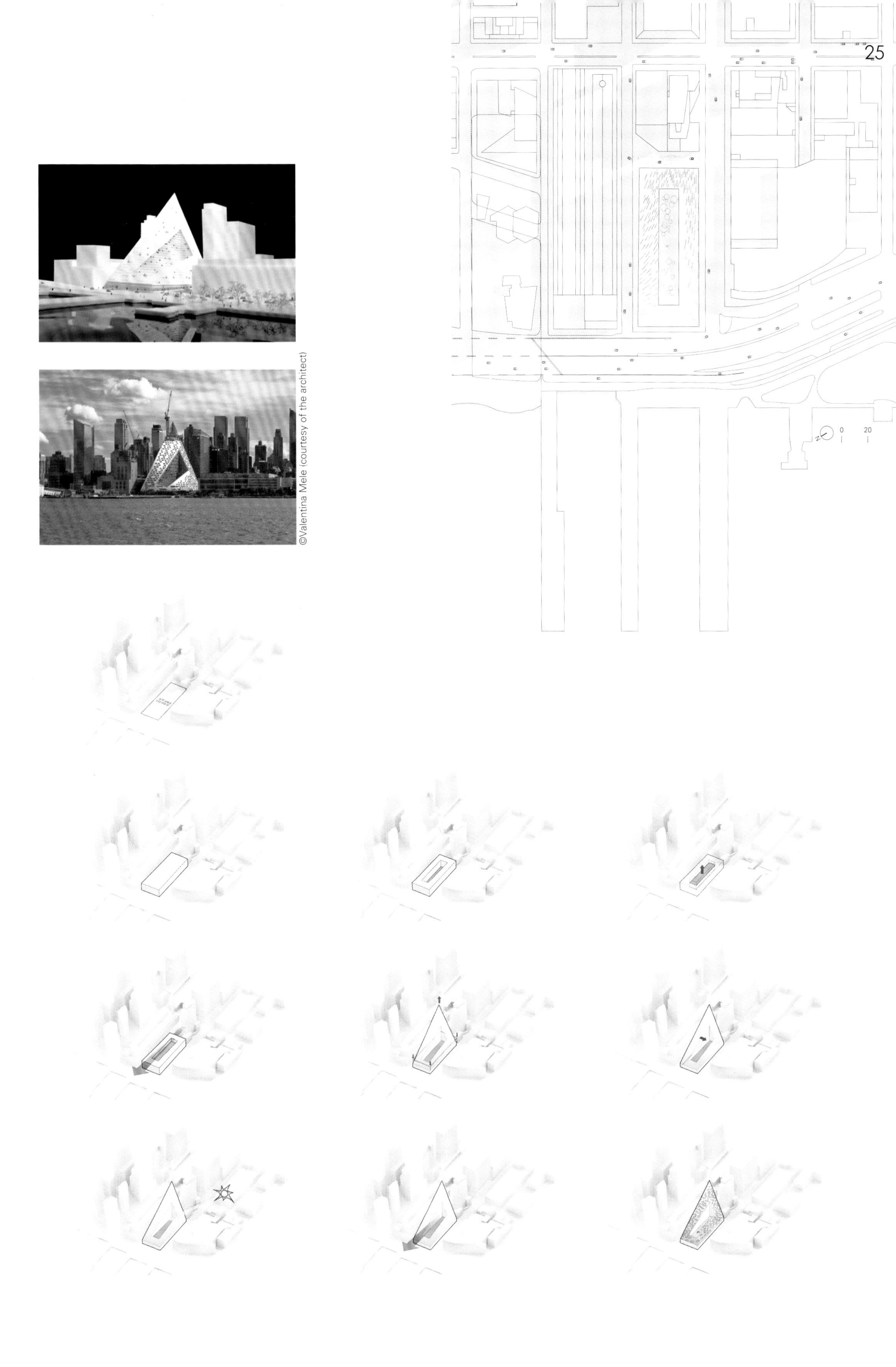

©Valentina Mele (courtesy of the architect)

courtyard building with the airiness and the expansive views of a skyscraper. By keeping three corners of the block low and lifting the north-east corner up towards its 450 ft peak, the courtyard opens views towards the Hudson River, bringing low western sun deep into the block and graciously preserving the adjacent Helena Tower's views of the river. The form of the building shifts depending on the viewer's vantage point. While appearing like a pyramid from the West Side Highway, it turns into a dramatic glass spire from West 58th Street. The courtyard which is inspired by the classic Copenhagen urban oasis can be seen from the street and serves to extend the adjacent greenery of the Hudson River Park into VIA. The slope of the building allows for a transition in scale between the low-rise structures to the south and the high-rise residential towers to the north and west of the site. The highly visible sloping roof consists of a simple ruled surface perforated by terraces - each one unique and south-facing. The fishbone pattern of the walls is also reflected in its elevations. Every apartment gets a bay window to amplify the benefits of the generous view and balconies which encourage interaction between residents and passers-by.

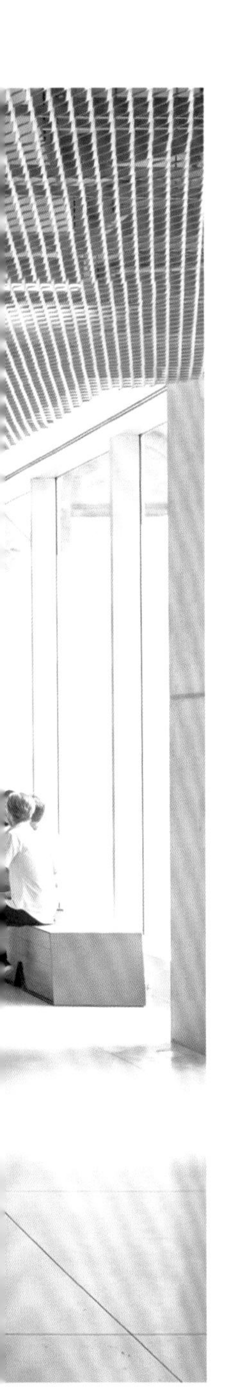

4 FEET 2 INCHES DEEP

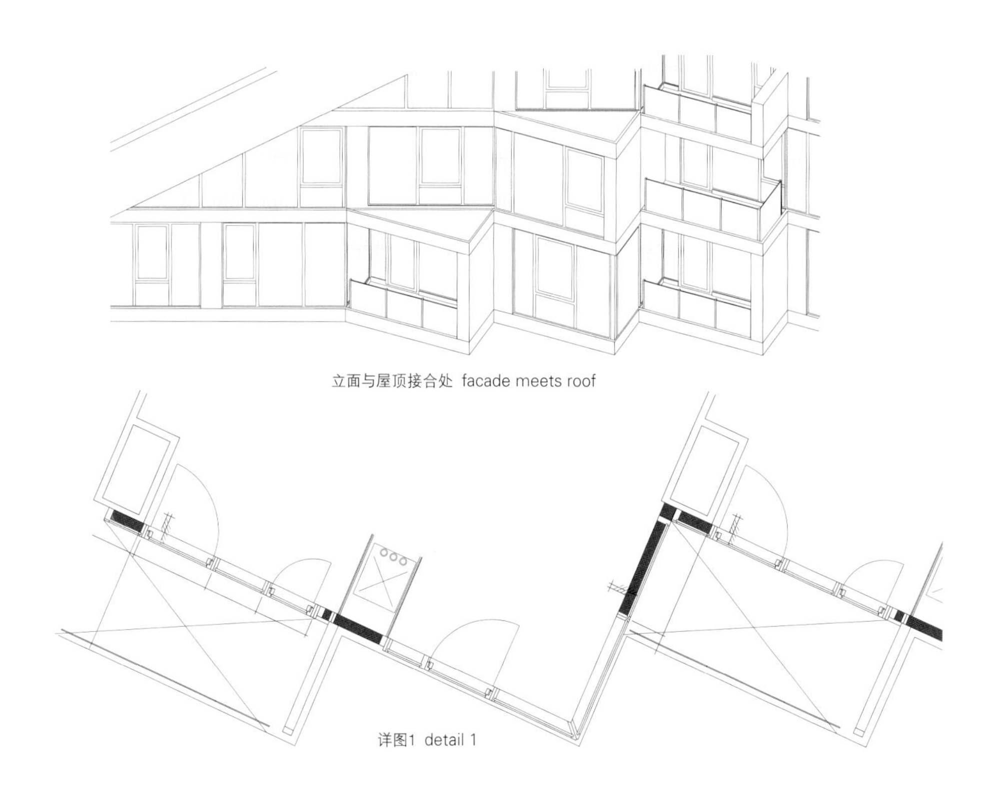

立面与屋顶接合处 facade meets roof

详图1 detail 1

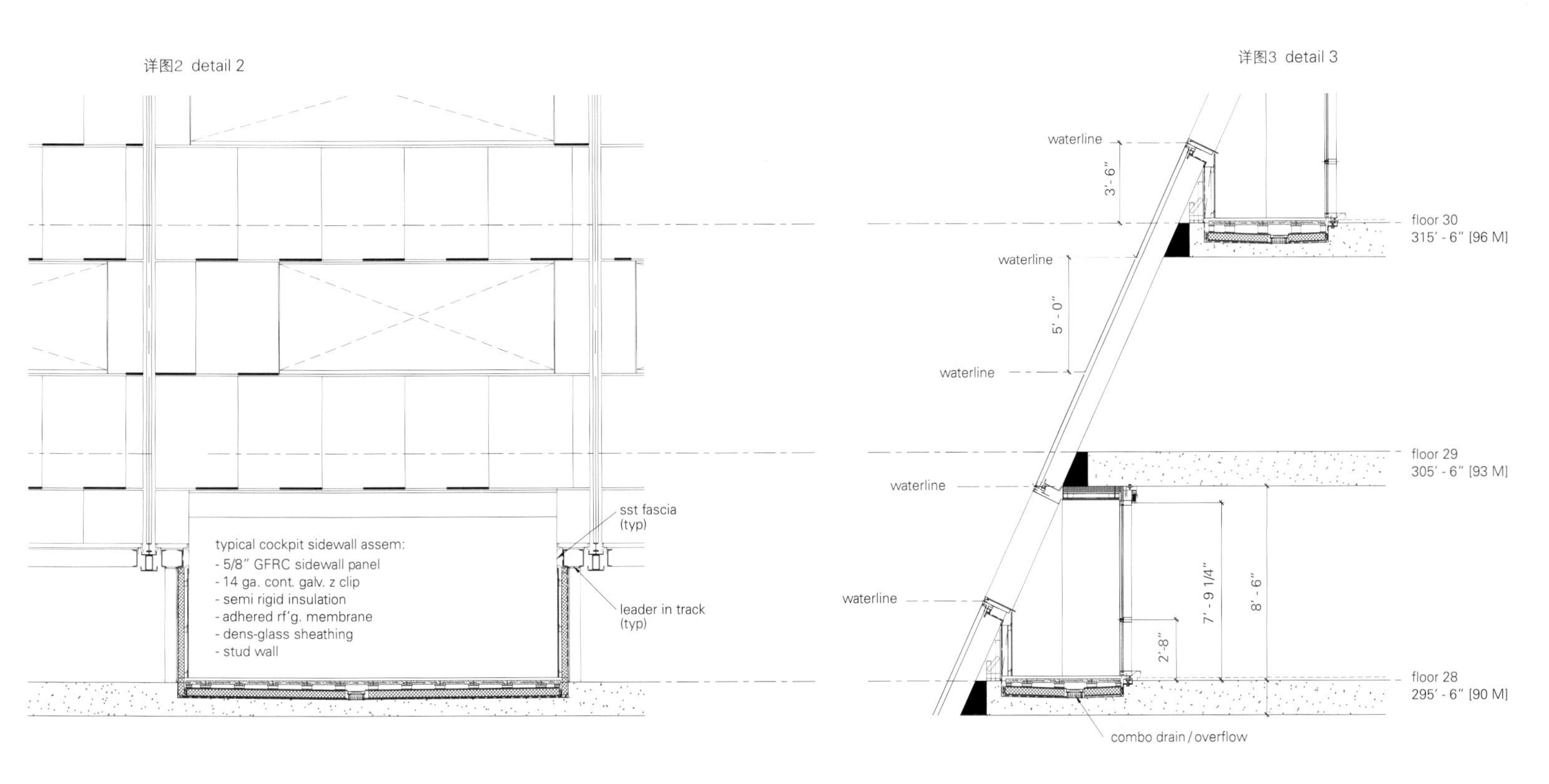

详图2 detail 2
typical cockpit sidewall assem:
- 5/8" GFRC sidewall panel
- 14 ga. cont. galv. z clip
- semi rigid insulation
- adhered rf'g. membrane
- dens-glass sheathing
- stud wall
sst fascia (typ)
leader in track (typ)
详图3 detail 3
waterline
3'-6"
waterline
5'-0"
waterline
waterline
waterline
floor 30
315' - 6" [96 M]
floor 29
305' - 6" [93 M]
7'-9 1/4"
8'-6"
2'-8"
floor 28
295' - 6" [90 M]
combo drain / overflow

项目名称：VIA 57 West / 地点：New York, USA / 建筑师：BIG
主管合伙人：Bjarke Ingels, Thomas Christoffersen, Beat Schenk / 项目建筑师：David Brown
项目团队：Aleksander Tokarz, Alessandro Ronfini, Alessio Valmori, Alvaro Mendive, Benjamin Schulte, Birk Daugaard, Celine Jeanne, Christoffer Gotfredsen, Daniel Sundlin, Dominyka Mineikyte, Eivor Davidsen, Felicia Guldberg, Florian Oberschneider, Soren Grunert, Gabrielle Nadeau, Gül Ertekin, Ho Kyung Lee, Hongyi Jin, Julian Liang, Julianne Gola, Laura Youf, Lucian Racovitan, Marcella Martinez, Maria Nikolova, Maya Shopova, Mitesh Dixit, Nicklas A. Rasch, Ola Hariri, Riccardo Mariano, Steffan Heath, Stanley Lung, Tara Hagan, Thilani Rajarathna, Tyler Polich, Valentina Mele, Valerie Lechene, Xu Li, Yi Li
项目负责人&室内设计：David Brown / 项目经理&室内设计：Beat Schenk
室内设计团队：Aaron Hales, Alessandro Ronfini, Brian Foster, Christoffer Gotfredsen, Ho Kyung Lee, Hongyi Jin, Ivy Hume, Jenny Chang, Lauren Turner, Mina Rafiee, Rakel Karlsdottir, Tara Hagan, Thomas Fagan, Tiago Barros, Valentina Mele
合作方：SLCE Architects, Starr Whitehouse, Thornton Tomasetti, Dagher Engineering, Langan Engineering, Hunter Roberts, Enclos, Philip Habib & Assoc, Vidaris Inc, Nancy Packes, Van Deusen & Assoc, Cerami & Assoc, CPP, AKRF, Glessner Group, Brandston Partnership Inc
客户：The Durst Organization
摄影师：courtesy of the architect-p.24, p.29
©Iwan Baan (courtesy of the architect)-p.20~21, p.23, p.26~27, p.28, p.32~33

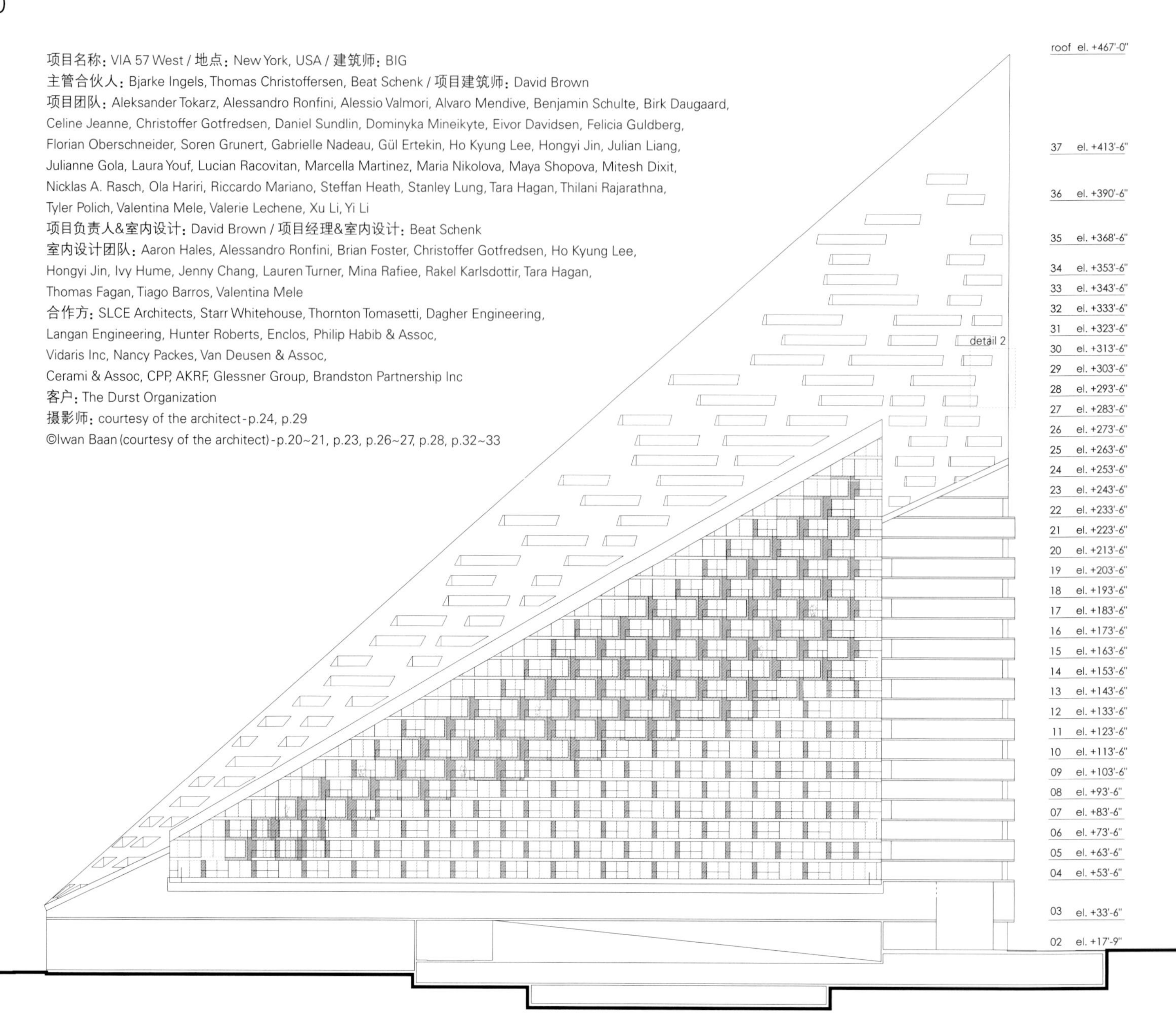

A-A' 剖面图 section A-A'

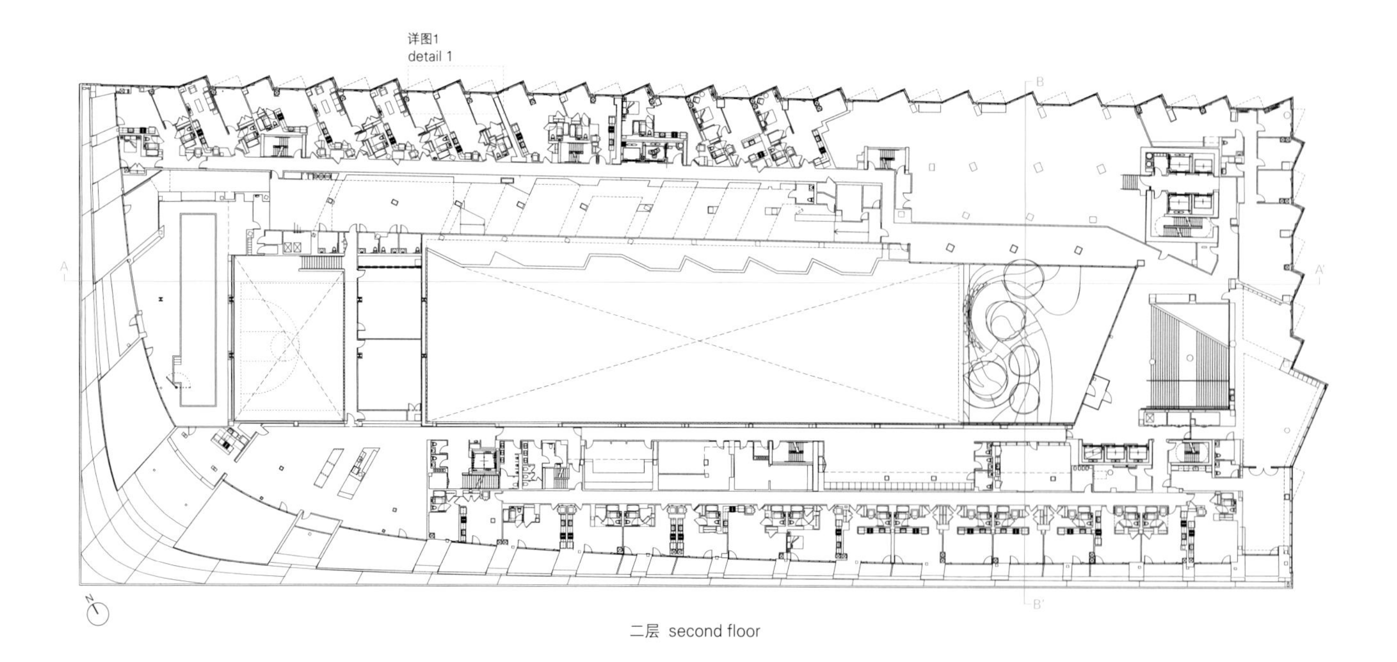

二层 second floor

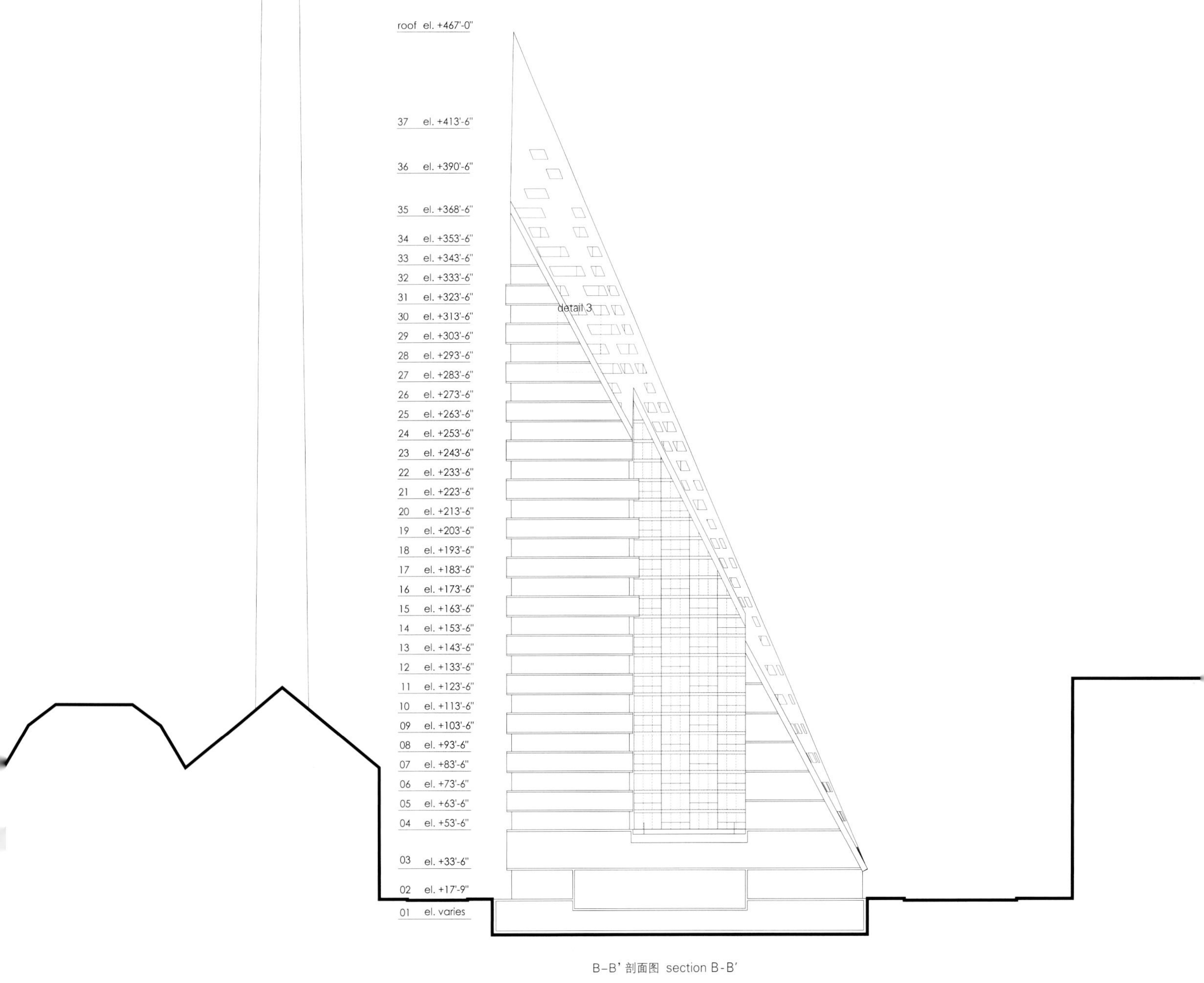

B–B' 剖面图 section B-B'

用途：housing / 用地面积：14,926.36m² 总建筑面积：77,202m² (商用：4,533.67m², 社区设施：2,561.12m², 庭院：2,006.70m², 住宅：75,382.46m², 便民设施：3,901.93m², 停车场：4,418m²) / 竣工时间：2016

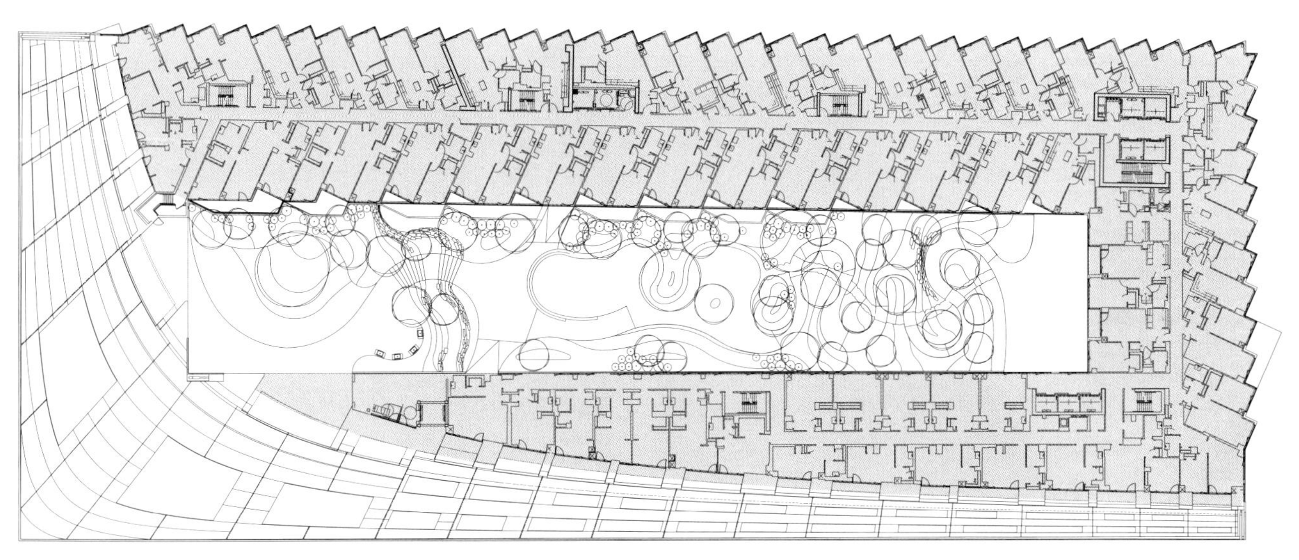

四层 fourth floor

伦纳德大街 56 号公寓

Herzog & de Meuron

NO PARKIN
8AM-6
MON
cite

摩天大楼是当代城市中的重要组成部分。定义摩天大楼这一建筑类型的唯一标准就是它的高度，然而如今却成了最没有个性的特征。典型的住宅大楼虽然成功地聚合了居住单元，但却往往不能改善居民的生活环境。虽然住宅单元达到了惊人的密度，但它们通常是在简单的、受到挤压的形式内重复叠加的单元，这种重复产生了既无额外优势也无建筑品质的、重复的、缺乏个性特征的构筑物。而对于生活在此种建筑中的人，这种雷同和重复的体验不能使人愉悦。伦纳德大街56号公寓反对这种无特征和重复性，它从以往这些众多的摩天大楼中脱颖而出。尽管它是大型建筑，但它的目的是实现一种独特甚至是私密的建筑品质。

该项目被设想成由单个房间堆叠而成的构造，每个房间都是独一无二的，让人能够从整体的堆叠构造中辨别出来。在仔细研究了当地的施工方法之后，建筑师发现了移动和改变楼板形成转角、悬臂和阳台的设计可能，这些都是很受欢迎的设计策略，可以为每间公寓都提供独特的居住条件。在大楼的底座部分，堆叠的体块呼应了街道的规模和当地的特殊条件，而大楼的顶部交错变化，与天空融为一体。大楼的中间楼层的交错变化较为收敛，变化更加细微，好似位于柱身当中。为了打破高层建筑重复而毫无特征的趋势，伦纳德大街56号公寓项目由室内向室外进行开发。项目从单个的房间开始，将它们视为"玻璃模块"，逐层地堆叠在一起。这些模块聚集在一起，直观地显示了大楼的体量并且塑造了大楼的外形。这些模块给人带来的内部体验就好似步入一系列巨大的飘窗当中。

从建筑的剖面也能看出"堆叠"房间的设计策略，这种策略形成了大量的露台和突出的阳台。在小心翼翼地避免直接俯瞰相邻公寓的同时，这些户外空间也提供了建筑物中人与人（或许是陌生人）之间间接的视觉联系。这些空中住宅聚集在一起，形成了一种有凝聚力的层叠关系，一个垂直的社区，有些类似于纽约的特殊街区，这些街区别具一格地融合了同等程度的亲密度和隐私性。

所有大楼的顶端都是它最醒目的元素，同样，伦纳德大街56号公寓的顶部也是该项目中最具表现力的部分，并且融入了纽约市传统的标志性大楼顶部的设计。楼顶的这种表现力直接受到室内空间需求的驱使，它包括十个具备宽敞户外空间和起居区的大型顶层公寓。这些大型的功能组件在外部显现为大尺度的体块，根据内部的配置以及在观景方面的特别要求悬挑和移位，最终形成了具有雕塑感的屋顶表现方式。同时，大楼的底座部分呼应了特里贝克地区的特性。这里是纽约的一部分，其特点是融合了多种不同的建筑尺度——从小型联排别墅到大型工业建筑以及市中心无处不在的高层建筑。通过将各种尺寸的"模块"组合在一起，包括一楼大堂、停车场和住房配套设施，这座大楼反映了每一座附近建筑的规模，并与之和谐相融。

大楼的整体外观是建筑师接受并优化当地简单而常见的施工方法的成果。作为一种体量，在结构允许的范围内，该建筑有着极端的设计比例，而且由于它的占地面积较小，显得格外高大而纤细。该建筑还展示了其结构"骨骼"，并且没有掩盖其覆层下方的建造方法。相反，裸露的水平混凝土楼板展示了施工过程中的逐层堆叠，而现浇裸露混凝土柱使人们可以从内部体验到项目的结构受力尺度。在每隔一两个楼层的立面单元上安装活动窗口，这样能使建筑的交错和堆叠系统变得更有生气。高层建筑的此种不寻常的特征也让住户可以直接控制新鲜空气的进入。

综合这些不同的高层建筑设计策略，譬如，由室内向室外、呼应当地尺度以及最大限度地发挥当地施工系统的潜力，产生了一栋由145间公寓组成的大楼，其中只有5间公寓是重复的。它给那些将在这里生活的人们一个独特的家，在整体堆叠的体块中享受有个性化的独特时光。

56 Leonard Street

The high-rise tower is an important ingredient within the contemporary city. However, towers have come to be defined solely by their height and, as a type, they have become anonymous. Typical residential towers, while successful in aggregating the living unit, often fail to improve upon the living environment. The multiplication of units within simple extruded shapes produces repetitive and anonymous structures with no extra benefits or architectural qualities despite the incredible densities they achieve. For those who live in these structures, this experience of sameness and repetition can be relatively unpleasant. 56 Leonard Street acts against this anonymity and repetitiveness, emanating from so many towers of the recent past. Its ambition is to achieve, despite its size, a character that is individual and personal, perhaps even intimate.

The project is conceived as a stack of individual rooms, where each room is unique and identifiable within the overall stack. A careful investigation of local construction methods revealed the possibility of shifting and varying floor-slabs to create corners, cantilevers and balconies – all welcome strategies for providing individual and different conditions in each apartment. At the base of the tower, the stack reacts to the scale and specific local conditions on the street, while the top staggers and shifts to engage with the sky. In-between, the staggering and variation in the middle-levels is more controlled and subtle, like in a column shaft. To break up the tendency towards repetition and anonymity in high-rise buildings, 56 Leonard Street was developed from the inside out. The project began with individual rooms, treating them as "blocks of glass" grouped together on a floor-by-floor basis. These blocks come together to directly inform the volume and to shape the outside of the tower. From the interior the experience of these blocks is like stepping into a series of large bay windows.

The strategy of stacking rooms also happens in section, creating a large number of terraces and projecting balconies. While being careful to avoid directly overlooking a neighboring apartment, these outdoor spaces provide indirect visual

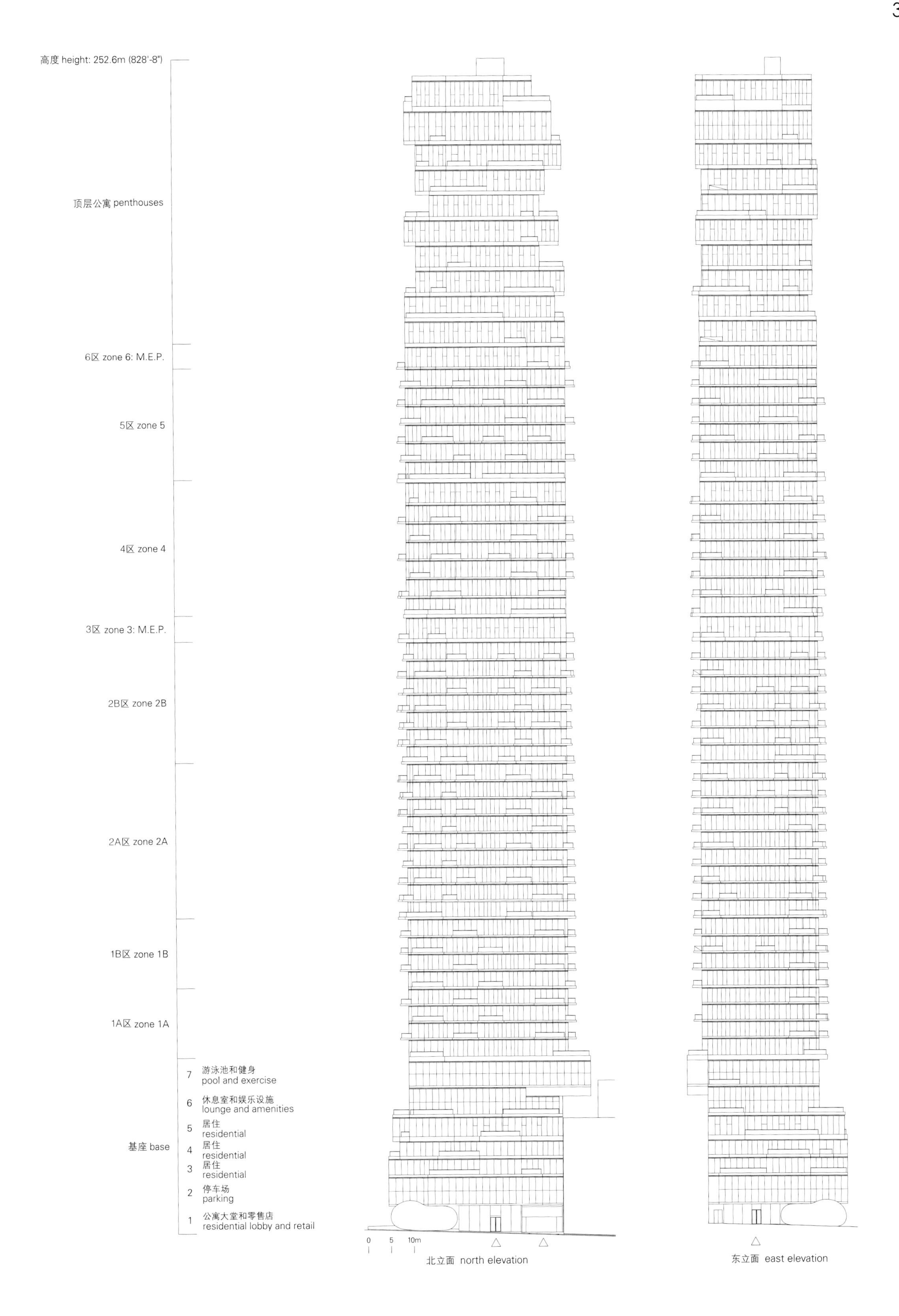
高度 height: 252.6m (828'-8")
顶层公寓 penthouses
6区 zone 6: M.E.P.
5区 zone 5
4区 zone 4
3区 zone 3: M.E.P.
2B区 zone 2B
2A区 zone 2A
1B区 zone 1B
1A区 zone 1A
7 游泳池和健身 pool and exercise
6 休息室和娱乐设施 lounge and amenities
5 居住 residential
基座 base
4 居住 residential
3 居住 residential
2 停车场 parking
1 公寓大堂和零售店 residential lobby and retail
0 5 10m
北立面 north elevation
东立面 east elevation

1. 门厅 2. 大型公共房间 3. 厨房 4. 露台 5. 主卧 6. 主浴室 7. 卧室 8. 浴室
9. 化妆室 10. 杂物间 11. 走廊 12. 开关设备室 13. 滞洪池 14. 警卫室

1. foyer 2. great room 3. kitchen 4. terrace 5. master bedroom
6. master bathroom 7. bedroom 8. bathroom 9. powder room 10. utility room
11. corridor 12. switch gear room 13. storm water detention 14. security room

四层_基座 third floor_base

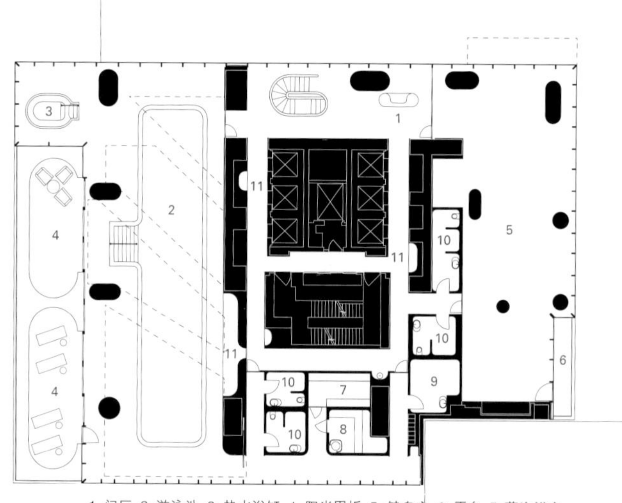

1. 门厅 2. 游泳池 3. 热水浴缸 4. 阳光甲板 5. 健身室 6. 露台 7. 蒸汽浴室
8. 桑拿房 9. 治疗室 10. 更衣室 11. 长椅

1. foyer 2. pool 3. hot tub 4. sun deck 5. exercise room 6. terrace
7. steam room 8. sauna 9. treatment room 10. changing room 11. bench

八层_基座 seventh floor_base

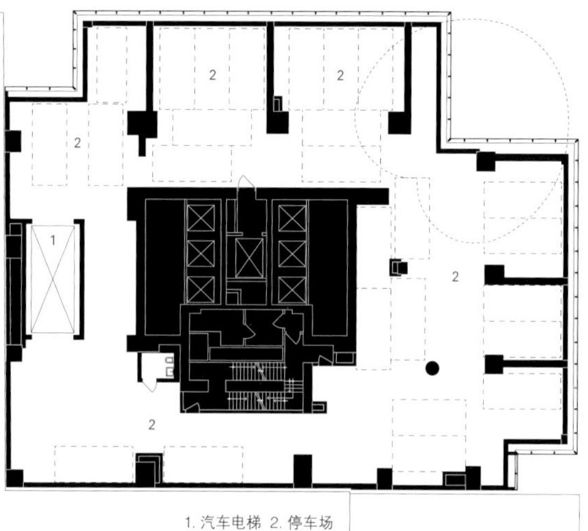

1. 汽车电梯 2. 停车场
1. car lift 2. parking

三层_基座 second floor_base

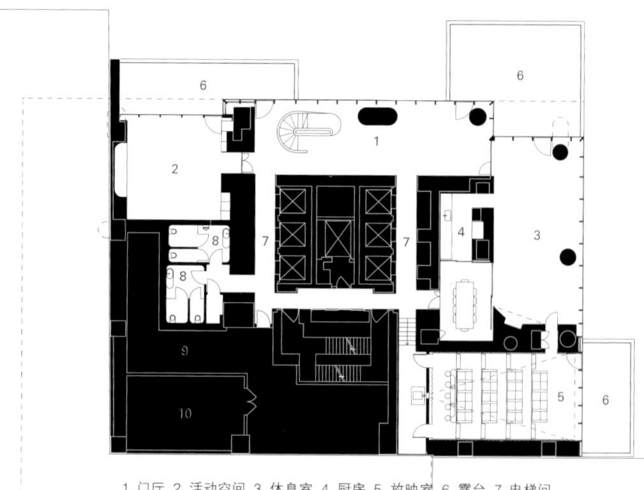

1. 门厅 2. 活动空间 3. 休息室 4. 厨房 5. 放映室 6. 露台 7. 电梯间
8. 卫生间 9. 机械与泳池设备 10. 便利发动机室

1. foyer 2. activity space 3. lounge 4. kitchen
5. screening room 6. terrace 7. elevator alcove 8. restroom
9. mechanical and pool equipment 10. convenience generator room

七层_基座 sixth floor_base

N 0 5 10m

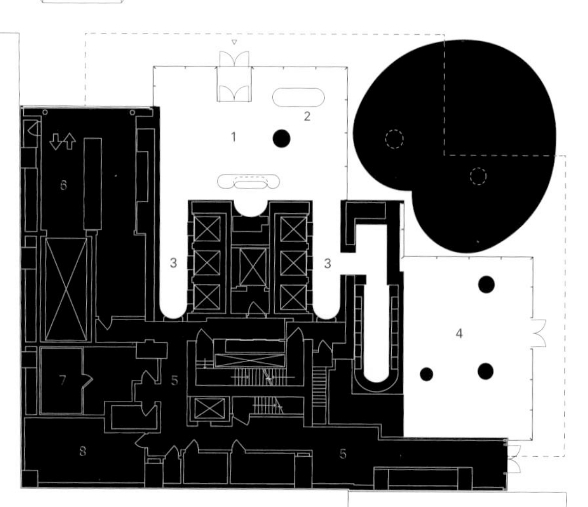

1. 公寓大堂 2. 座位区 3. 电梯间 4. 零售区 5. 走廊 6. 停车场入口 7. 打包室
1. residential lobby 2. seating area 3. elevator alcove
4. retail area 5. corridor 6. parking garage entry 7. package room

一层_基座 ground floor_base

1. 门厅 2. 大型公共房间 3.厨房 4. 露台 5. 主卧 6. 主浴室 7. 卧室 8. 浴室
9. 化妆室 10. 杂物间 11. 走廊 12. 早餐室 13. 气体增压室

1. foyer 2. great room 3. kitchen 4. terrace 5. master bedroom
6. master bathroom 7. bedroom 8. bathroom 9. powder room
10. utility room 11. corridor 12. breakfast room 13. gas booster room

六层_基座 fifth floor_base

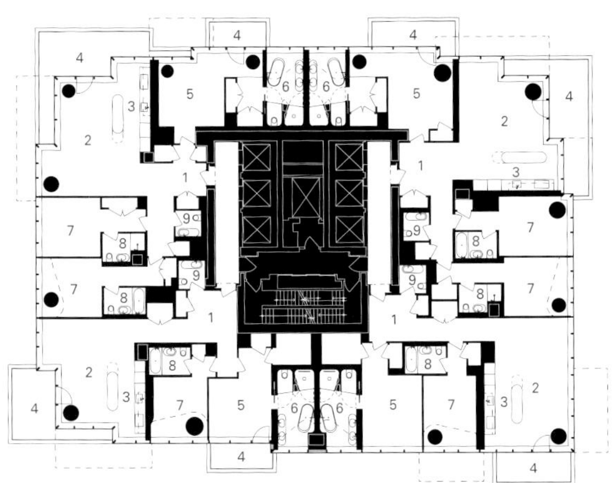

1. 门厅 2. 大型公共房间 3. 厨房 4. 露台 5. 主卧 6. 主浴室 7. 卧室 8. 浴室 9. 化妆室
1. foyer 2. great room 3. kitchen 4. terrace 5. master bedroom
6. master bathroom 7. bedroom 8. bathroom 9. powder room
26层_2B区 twenty fifth floor_zone 2B

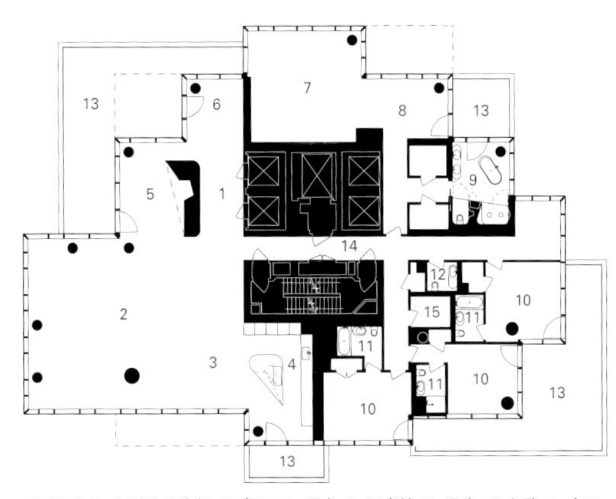

1. 门厅 2. 大型公共房间 3. 餐厅 4. 厨房 5. 图书馆 6. 温室 7. 主卧 8. 客厅
9. 主浴室 10. 卧室 11. 浴室 12. 化妆室 13. 露台 14. 走廊 15. 杂物间
1. foyer 2. great room 3. dining 4. kitchen 5. library 6. conservatory
7. master bedroom 8. sitting room 9. master bathroom 10. bedroom
11. bathroom 12. powder room 13. terrace 14 corridor 15. utility room
50层_7区 forty ninth floor_zone 7

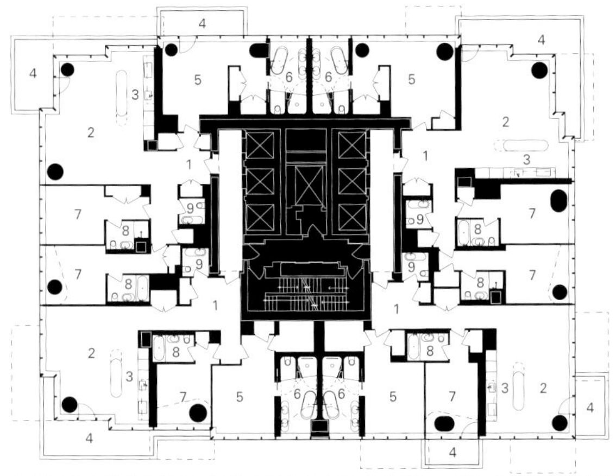

1. 门厅 2. 大型公共房间 3. 厨房 4. 露台 5. 主卧 6. 主浴室 7. 卧室 8. 浴室 9. 化妆室
1. foyer 2. great room 3. kitchen 4. terrace 5. master bedroom
6. master bathroom 7. bedroom 8. bathroom 9. powder room
17层_2A区 sixteenth floor_zone 2A

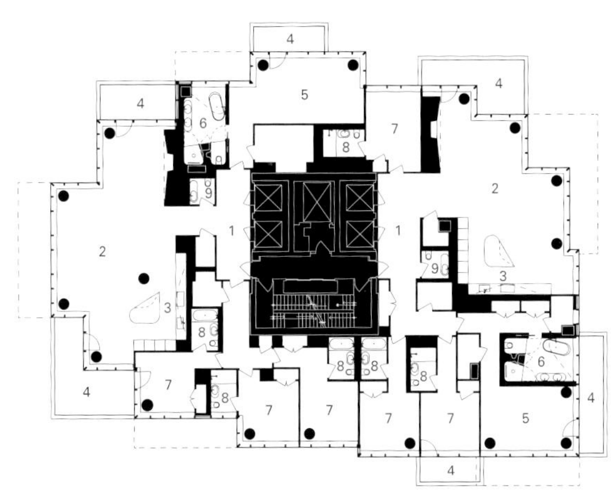

1. 门厅 2. 大型公共房间 3. 厨房 4. 露台 5. 主卧 6. 主浴室 7. 卧室 8. 浴室 9. 化妆室
1. foyer 2. great room 3. kitchen 4. terrace 5. master bedroom
6. master bathroom 7. bedroom 8. bathroom 9. powder room
43层_5区 forty second floor_zone 5

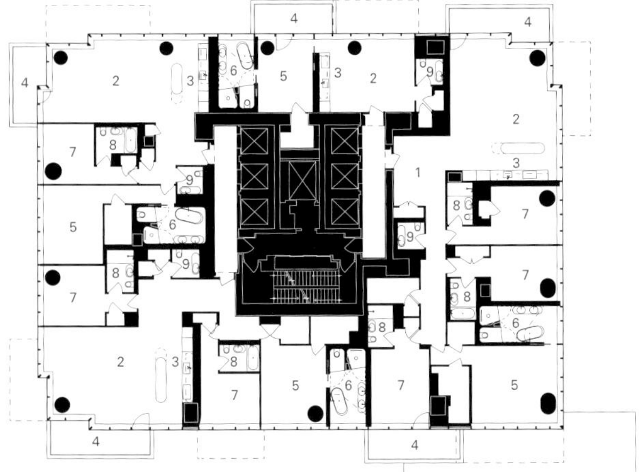

1. 门厅 2. 大型公共房间 3. 厨房 4. 露台 5. 主卧 6. 主浴室 7. 卧室 8. 浴室 9. 化妆室
1. foyer 2. great room 3. kitchen 4. terrace 5. master bedroom
6. master bathroom 7. bedroom 8. bathroom 9. powder room
13层_1B区 twelfth floor_zone 1B

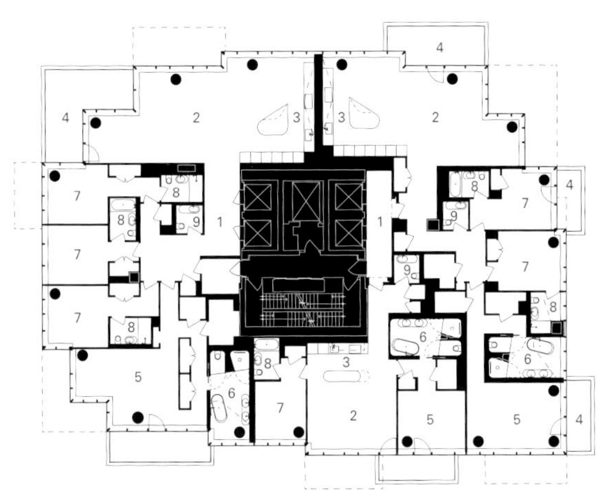

1. 门厅 2. 大型公共房间 3. 厨房 4. 露台 5. 主卧 6. 主浴室 7. 卧室 8. 浴室 9. 化妆室
1. foyer 2. great room 3. kitchen 4. terrace 5. master bedroom
6. master bathroom 7. bedroom 8. bathroom 9. powder room
35层_4区 thirty fourth floor_zone 4

项目名称：56 Leonard Street / 地点：New York, NY, USA / 建筑师：Herzog & de Meuron / 合伙人：Jacques Herzog, Pierre de Meuron, Ascan Mergenthaler_ Partner in charge / 项目团队：Philip Schmerbeck, Associate, Project director; Mehmet Noyan, Associate, Project manager; Vladimir Pajkic, Associate; Caroline Alsup, Mark Chan, Simon Filler, Dara Huang, Sara Jacinto, Jin Tack Lim, Mark Loughnan, Jaroslav Mach, Donald Mak (Associate), Hugo Miguel Moura, Jeremy Purcell, Chantal Reichenbach, James Richards, Heeri Song, Charles Stone, Kai Strehlke, Zachary Martin Vourlas, Jason Whiteley, Sung Goo Yang, Daniela Zimmer, Christian Zöllner, Iwona Boguslawska, Martin Raub, Josh Helin, Joem Elias Sanez / 公共艺术品创作：Anish Kapoor, London / 设计顾问/室内设计：Herzog & de Meuron / 执行建筑师：Goldstein, Hill & West Architects, LLP / 施工管理：Lend Lease
机械工程：Cosentini Associates / 结构工程：WSP Cantur Seinuk / 客户：Alexico Group, New York / 用地面积：1,162m² / 建筑面积：1,162m² / 总建筑面积：45,291m² / 楼层数：57 / 长×宽×高：38 x 30 x 253m / 立面面积：30,472m² / 用途：Residential units; 39,718m² (146 residential units) / Residential amenities; 1,742m² (screening room, lap length swimming pool, gym, spa, lounge and activity space)
Residential lobby; 190m² (concierge, mailroom and lobby) / Public art; 167m² polished stainless steel sculpture in collaboration with Anish Kapoor / Residential parking; 1,811m² of residential parking at level 2 / Commercial retail storefront; 135m² of retail space at ground level storefront / 材料&制造商：windows; 4-sided structurally glazed extruded aluminum window wall by Enclos, operable windows by Schuco / Exposed structural concrete slab edges, terraces, and balconies, 8,600 psi, polished exposed circular perimeter columns, 12,000 psi. / Flooring in units; white stained solid appalachian white oak, rift and quartered, 3/4" thick, 3" or 4" wide. random length btw 3' to 10' (3'-0" to be ~10%) / Ceilings in units; flat white paint over kadex over concrete slab
概念设计时间：2006.10—2007.2 / 设计开发时间：2007.3—2007.12 / 施工文件制作时间：2007.12—2009.1 / 施工时间：2012.1—2017.6
摄影师：©Hufton + Crow (courtesy of the architect) - p.42~43, p.45, p.46~47; ©Iwan Baan (courtesy of the architect) - p.34~35, p.36, p.38, p.44

links between people – maybe strangers – who share the building. Aggregated together, these houses-in-the-sky, form a cohesive stack, a vertical neighborhood, somewhat akin to New York's specific neighborhoods with their distinctive mix of proximity and privacy in equal measure.

The top of any tower is its most visible element and, in keeping with this, the top of 56 Leonard Street is the most expressive part of the project and relates to the tradition of iconic tower tops in New York City. This expressiveness is driven directly by the requirements of the interior, consisting of ten large-scale penthouses with expansive outdoor spaces and spacious living areas. These large program components register on the exterior as large-scale blocks, cantilevering and shifting according to internal configurations and the desire to capture specific views that ultimately result in the sculptural expression of the top. Meanwhile, the base of the tower responds to the special character of Tribeca. This is a part of New York characterized by a wide range of building scales – from small townhouses to large industrial blocks and the ubiquitous high-rise buildings of downtown. By grouping together blocks of various sizes, including the lobby, the parking deck, housing amenities and a few apartments, the tower reflects and incorporates each of these neighborhood scales.

The overall appearance of the tower is very much a result of accepting and pushing to the limit simple and familiar local methods of construction. As a volume, the building has extreme proportions – at the very edge of what is structurally possible – and given its relatively small footprint, is exceptionally tall and slender. The building also shows its structural "bones" and does not hide the method of its fabrication underneath layers of cladding. Instead, exposed horizontal concrete slabs register the floor-by-floor stacking of the construction process and exposed in-situ concrete columns allow the scale of the structural forces at work to be experienced from within the interior. The system of staggering and stacking is further animated through operable windows in every second – or third – facade unit. This unusual feature for high-rise buildings also allows occupants to directly control fresh air intake.

Together these different strategies – considering the tower from the inside-out, responding to local scales, and maximizing the potential of local construction systems – produce a building where only five out of the 145 apartments are repeated, giving those who will live in this project their own unique home characterized by distinct moments of individuality within the overall stack.

复兴混凝土

A Concr

Renaiss

在近一个世纪的全球传播过程中，钢筋混凝土已经建立了一种稳固的思想传统。混凝土持续渗透至我们的生活当中。混凝土是一种随处可见的元素，它要么被主流艺术家和建筑师使用，要么只是我们日常接触到的一般构筑物和人工制品的一部分。正如艾德里安·福尔蒂所说的，因为混凝土极其柔韧，所以成为探索挑战传统建筑构造的形式词汇的首选媒介。露石混凝土墙就能展现出许多不同的纹理和颜色。通过或多或少有些复杂的模板系统，建筑师甚至可以将其他材料，如木料、金属板或砖块的纹理和颜色转移至混凝土表面上。事实上，混凝土表面非常适合营造简朴的氛围，也可以形成给人带来丰富感官享受的空间体验。钢筋混凝土结构可以以

In approximately one century of global dissemination, the reinforced concrete has built up a solid intellectual tradition. The concrete pervades our lives relentlessly. Either used by mainstream artists and architects or just part of our everyday contact with the most prosaic structures and artefacts, the concrete is all over the place. The concrete's extreme pliability, as Adrian Forty put it, makes it a preferred medium to explore formal vocabularies that challenge conventional building tectonics. Exposed concrete walls can be produced to show many different textures and colours. Even the textures and colours of other materials such as wood, metallic panels, or bricks, transferred onto its surfaces through more or less sophisticated formwork systems. In fact, concrete surfaces are ideal to create austere atmospheres but

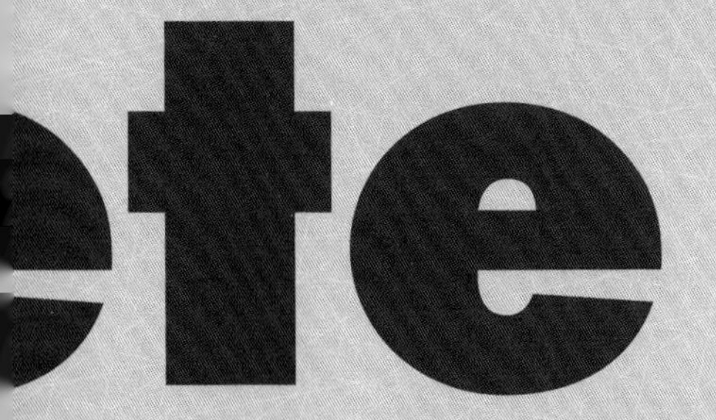

ance

萨维耶斯某住宅_House in Savièse / Anako'architecture
生态观测住宅_Ecoscopic House / archipelagos - Manolo Ufer
U Retreat度假屋_U Retreat / IDMM Architects
FU住宅_FU-House / Kubota Architect Atelier
上海西岸Fab-Union Space建筑_Fab-Union Space / Archi-Union Architects
Bosjes教堂_Bosjes Chapel / Steyn Studio
森林独奏屋_Solo House / Office KGDVS
FIM新总部_New FIM Headquarters / Localarchitecture

复兴混凝土_A Concrete Renaissance / Nelson Mota

传统建筑技术无法实现的方式来挑战地心引力。即使是在偏远地区，建筑师也可以将混凝土当成最简单的建筑系统。在许多地方，混凝土目前已成为新的地方传统中不可或缺的一部分。很难找到一种材料能够在截然不同的环境中表现出此种潜力。这种应对这些不同挑战的潜力已经重新燃起人们对混凝土作为建筑表现媒介的兴趣。在经过了混凝土被指责为一种与建筑现代主义阴暗面有关联的材料的这段时期之后，如今，我们可以见证到混凝土的复兴。

also to suggest voluptuous spatial experiences. Reinforced concrete structures can challenge gravity in ways that traditional building techniques could not. But they are also used as the most straightforward building system available even in remote locations. In many places, the concrete is now part and parcel of the new vernacular tradition. One can hardly find a material with such potential to perform in contrasting circumstances. This potential to cope with such different challenges has contributed to instigating a renewed interest in the concrete as a medium of architectural expression. After the period when it was stigmatized as a material associated with the darker side of architectural modernism, today we can witness a concrete renaissance.

生态观测住宅，墨西哥
Ecoscopic House, Mexico

U Retreat度假屋，韩国
U Retreat, Korea

1971年，在宾夕法尼亚大学路易·康的著名大师班上，他主张如果需要呈现某样东西，你就应该考虑其性质。并且他还主张设计知识是解码隐藏于性质中的经验教训的重要工具。他用一个例子说明了这一想法："例如，如果你想到砖头，就问它：'你想要什么，砖头？'而砖头回答说：'我想要一个拱门。'"砖头具备永恒的维度，因此这样的回答能够令人信服。如果我们关注一种历史较短的材料，例如，钢筋混凝土，会得到怎样的答案呢？

与砖头一样，钢筋混凝土也是人类想象力和技术的产物。虽然混凝土的历史要比砖头短得多，但在近一个世纪里，它在建筑环境以及设计学科和工程学的实践过程中引发了一场革命。混凝土被迅速地应用于建筑业当中，使得一时之间混凝土建筑随处可见。现在，如果我们想到钢筋混凝土并问道"你想要什么，混凝土？"我们会得到怎样的答案呢？我不知道康是否也思考过在此种情况下可能得到的答案。然而，在前面报道的项目中，我们意识到答案可能并不简单。

正如艾德里安·福尔蒂在对混凝土材料历史的华丽描述中所说的，即使是在20世纪末——混凝土经历了一个多世纪的全球传播之后——人们仍在努力思考这一问题。混凝土的文化同化尚未完成。[1]"在建筑学领域，"福尔蒂认为，"有一种思想传统将混凝土视为使建筑能够完成其使命的材料，使人们实现那些以往因缺乏途径而无法实现的梦想。"福尔蒂还认为，另一种传统与混凝土的新颖性有关，这种传统将混凝土这一媒介置于历史之外，"并为建筑学提供了摆脱过去负担的途径"。

In his famous masterclass at the University of Pennsylvania in 1971, Louis Kahn argued that when you need to give something presence you should consult the nature. Further, he contended that the design knowledge was a key instrument to decode nature's hidden lessons. He illustrated the idea with an example. "If you think of brick, for instance, you say to brick: 'What do you want, brick?' And brick says to you: 'I'd like an arch'." Brick has a timeless dimension that may allow for such a compelling statement. What if we focus on a material with less pedigree, the reinforced concrete, for example?
Like the brick, the reinforced concrete is a product of human imagination and technology. While it is relatively much younger than the brick, over the last century it has produced a revolution in the built environment, and in the practice of the design disciplines and engineering. It was adopted by the building industry so rapidly that concrete buildings were suddenly all over the place. Now, if we think of the reinforced concrete and we ask "What do you want, concrete?", what answer are we likely to get? I don't know whether Kahn has also thought about a possible answer for this case. However, in the projects featured ahead we realize that the answer could hardly be simple.
As Adrian Forty has put it in his magnificent account of concrete's material history, even at the end of the 20th century – more than a century after its global dissemination – people were still working on what to think about it. The concrete's cultural assimilation was not complete yet.[1] "Within the field of architecture", Forty argues, "there is one intellectual tradition that sees the concrete as having made it possible for architecture to fulfil its destiny, to achieve things that people in previous times had dreamed of, but had lacked the means to realize." The other tradition, according to Forty, was related with the concrete's novelty, which put this medium outside history "and offered architecture the means to break free from the burden of its past".
In the following projects, the concrete's ambivalent nature can be clearly observed. One can see the concrete being used

萨维耶斯某住宅，瑞士
House in Savièse, Switzerland

在接下来的项目中，我们可以清楚地看到混凝土矛盾的性质。我们可以看到混凝土作为一种媒介，建立了与历史之间的对话，而其他的设计案例则明显地想要表现混凝土的新颖性。既然如此，我们该如何辨别这些项目属于哪一种思想传统呢？我认为几何结构可能是回答这一问题的重要信号。正如路易·康所指出的，大多数材料都具有一种更偏向于自然的几何用途。但对于混凝土来说，情况好像并非如此。混凝土在适应不同几何结构方面的多功能性似乎是无限的。根据在后续章节中报道的项目，我将挑选出作为建筑媒介的混凝土在应用过程中的三种几何表现方式：倾斜、曲线和平面。

1. 倾斜功能

1966年，法国艺术家保罗·维利里奥出了一本名为《倾斜功能》[2]的小册子形式的宣言。在宣言中，他详细说明了一种"倾斜理论"。作为垂直墙体与水平表面的一种连接方式，倾斜建筑理念旨在批判历史上的建筑空间范型。而倾斜的平台将引发全新的亲身体验和一种新的空间感。"斜轴和倾斜的平面设计"，他主张，"实现了所有形成城市新秩序所需的必要条件，并且也允许全面重塑建筑词汇。"

由Archipelagos设计的生态观测住宅位于墨西哥蒙特雷的一处场地，该项目很好地展示了倾斜建筑的力量，以激发人们对空间的感受。生态观测这一设计理念很有意义。正如该项目的设计者所说的，"生态观测的理念并没有将建筑解读为一个（静态）物体，而是一个媒介

as a medium to create a dialogue with the history, and other cases that bluntly want to show their newness. Now, how can we identify which intellectual tradition do these projects belong to? I would argue that the geometry is a key indicator to produce a possible answer to this question. As Louis Kahn has pointed out, most materials have a preferred – natural? - geometric application. But with the concrete, things are seemingly different. The concrete's polyvalence in adapting to different geometric configurations seems boundless. Considering the projects featured in the following section, I will single out three geometric expressions in the use of concrete as an architectural medium: The oblique, the curve, and the plan.

1. The Oblique Function

In 1966 the French artist Paul Virilio published a small manifesto called "*The Oblique Function*".[2] In this piece he has elaborated a "theory of the oblique". The concept of oblique architecture was meant as a critique of the historical paradigm of architectural space as an articulation of horizontal surfaces with vertical walls. The oblique platform, instead, would trigger new bodily experiences and a new sense of space. "The oblique axis and the inclined plan", he argued, "realize all the necessary conditions for the creation of a new urban order and permit as well a total reinvention of the architectural vocabulary."

The Ecoscopic House, designed by Archipelagos for a site located in Monterrey, Mexico is a good demonstration of the power of oblique architecture to activate one's perception of space. The concept of ecoscope is meaningful. As the authors of the project put it, "The ecoscope is a concept by which the architecture is understood not as an object (static), but as a medium (interface) through which man (re-) defines the relationship with his surroundings." While in the Ecoscopic

尼泰罗伊当代艺术博物馆，奥斯卡·尼迈耶，巴西，1996年
Niterói Contemporary Art Museum, Oscar Niemeyer, Brazil, 1996

（界面），人们通过它可以（重新）定义与周围环境的关系。”位于蒙特雷的生态观测住宅项目采用了平整的楼面，精心设计的倾斜墙壁、窗户、楼梯、天窗等之间的相互作用激发了住户的空间体验。在该住宅项目中，混凝土的应用在激活多个相互作用的能量流方面起到了重要的作用。混凝土在这里担任了多重角色：结构、保温、饰面。项目设计者认为，正是混凝土卓越的可塑性使它成了使生态观测住宅成为“一个其主题具有重要意义的框架”的关键元素。

在瑞士事务所Anako' architecture为萨维耶斯（瑞士）设计的住宅项目中，倾斜功能同样发挥了重要的作用。混凝土建筑体量的几何外形是通过顺利地复制场地地形的方式而设计出来的。倾斜的线条定义了房屋的构造（从剖面和平面设计上），并激发了人们对空间的感知。经过精心打造，裸露的混凝土体量形成了一系列的空间和平台，定义了内部和外部、家庭空间和景观之间模糊的边界。

U Retreat度假屋由Heesoo Kwak和IDMM建筑师事务所设计，位于洪川郡（韩国）一处风景优美的场地，该项目结合了上面讨论过的两个项目的特质。一方面，建筑构件的组成激发了全新的空间体验。建筑体量和自然之间的关系成为项目关注的重点。项目试图回答这样的问题：“建筑如何体现季节和气候，以及每个瞬间都在变化的悬崖峭壁的运动？”可以说，混凝土和倾斜功能的使用便是该问题的答案。“混凝土体量形成的倾斜线条，”设计者认为，“能够与大自然多变的不规则线条形成对比。”也就是说，混凝土的使用不仅能形成意义深远的对比效果，同时还能够形成建筑物和大自然之间的一种共生组合。

House in Monterrey the floors are flat, and the dweller's experience of the space is activated by a carefully curated interaction between oblique walls, windows, stairs, skylights and so on. The use of concrete plays a key role in the activation of the multiple flows of energy that interact in the house. The concrete is used to perform several roles: Structure, insulation, and finishing. It is concrete's remarkable malleability that makes it a crucial element to make the Ecoscopic House "a frame within which the subject matter is relevant", as the authors contend.

In the house designed by the Swiss office Anako'architecture for Savièse (Switzerland), the oblique function performs also an important role. The concrete volume is shaped in such a way that its geometry smoothly replicates that of the site's topography. The oblique lines that define the house's composition – in the section as well as in the plan – stimulate the sensorial perception of space. The exposed concrete volume is carefully crafted to produce a sequence of spaces and platforms that define ambiguous borders between interior and exterior, between the domestic space and the landscape.

The project U Retreat, designed by Heesoo Kwak and IDMM Architects for a scenic location in Hongcheon-gun (Korea) combines the qualities of the two projects discussed above. On the one hand the composition of the architectural elements stimulates new spatial experiences. The relationship between the built mass and nature is a major concern of the project. The project tries to answer the question: "How could architecture embody the seasons and the climate, as well as the movement and the flow of the cliff which change with each and every instant?" The use of concrete and the use of the oblique function, as it were, were the answers to this question. "Oblique lines made by the concrete mass," the authors contend, "can contrast with irregular and varied lines of the nature." In other words, the concrete is used to simultaneously create meaningful contrasts and make a symbiotic combination between the architectural object and nature.

森林独奏屋，西班牙
Solo House, Spain

Bosjes教堂，南非
Bosjes Chapel, South Africa

2. 失重的曲线

在奥斯卡·尼迈耶的回忆录《时间曲线》（2007）的引语中，他宣称："直角或是人造的强硬刻板的直线都不能吸引我，能够吸引我的是自由而感性的曲线。这些曲线刻画在祖国的山脉上、蜿蜒在河流中、随着大海的波浪翻滚，还有你深爱之人身体的曲线，整个宇宙都是由曲线构成的，这就是爱因斯坦的曲线宇宙。"[3]所以尼迈耶喜欢将混凝土用作建筑表达的媒介之一并非偶然。他将自己对曲线的热情与混凝土表现出的可塑、结构和感官可能性很好地结合在了一起。在本章节介绍的项目中，建筑师还探索了一些其他的可能性。

上海西岸Fab-Union Space建筑项目就是一个例子。它是一家由上海创盟国际建筑设计有限公司设计的艺术馆，位于中国上海一处紧凑的场地上。在该项目中，建筑师利用混凝土的潜力创建了一栋软硬表面对比组合的建筑。定义建筑外壳的刚性体量与由弯曲墙壁组成的起伏的交通核心并存。这些墙体创造了持续的楼层过渡，而且也是建筑师用以激发艺术馆充满活力的用途的主要构件。

在Fab-Union Space艺术馆项目中，曲线与运动有关，而在位于南非维岑堡市一个由斯泰恩工作室设计的Bosjes教堂项目中，建筑师将曲线用作一种工具，围绕着一个单独的开放空间（礼堂）创造了一个仿佛无边无际的雕塑式外壳。建筑形式上的品质所传达的"诗意的运动"暗示了建筑内部精神体验的亲密性与周围景观超凡脱俗的品质之间的视觉交流。与尼迈耶设计的许多项目（例如，尼泰罗伊当代艺术博物馆）一样，在这个受《诗篇36：7》启发而设计的教堂项目中，混凝土的使用实现了垂直和水平构件的融合。楼板、墙面和屋顶成为连续构件

2. Weightless Curves

In the epigraph of Oscar Niemeyer's book of memoirs, *The Curves of Time*(2007), he declares: "It is not the right angle that attracts me, nor the hard, inflexible straight line, man-made. What attracts me are free and sensual curves. The curves in my country's mountains, in the sinuous flow of its rivers, in the waves of the sea, in the beloved woman's body. The whole universe is made of curves, Einstein's curve universe."[3] It is not casual that the concrete was also one of Niemeyer's favourite media for architectural expression. There was a happy marriage between his passion for curves and the plastic, structural and sensorial possibilities conveyed by the concrete. These were also some of the possibilities explored in the projects featured in this section.

The project for the Fab-Union Space, a gallery designed by Archi-Union Architects for a compact plot in Shanghai (China) is a case in point. In this project, the architects take advantage of the concrete's potential to create a building with a contrasting combination of hard and soft surfaces. The rigid volume that defines the shell of the building is juxtaposed with an undulating circulation core made of sinuous walls. These walls create continuous level transitions and are the main elements used by the architects to stimulate a dynamic use of the gallery.

While in the Fab-Union Space gallery the curves are associated with motion, in the Bosjes Chapel, designed by Steyn Studio for a location in Witzenberg (South Africa), the curves are instrumental to create a boundless sculptural shell around one single open space: the assembly hall. The "poetic motion" conveyed by the building's formal qualities suggests a simultaneous visual exchange between the intimacy of the spiritual experience inside the building, and the otherworldly qualities of the landscape surrounding it. As in many projects designed by Niemeyer – the Niterói Contemporary Art

FU住宅，日本
FU-House, Japan

的一部分，挑战了重力并消除了建筑体量和周围景观之间的界限。

在上面讨论的两个项目中，曲线与复杂的三维几何结构有关，从而产生了强有力的形式表述方式。然而，混凝土产生富有表现力的体量和空间的潜力也可以通过最基本的二维曲线形状——圆形得以实现。由瑞士事务所Localarchitecture设计的FIM（国际摩托车联盟）新总部以及由比利时事务所KGDVS设计的森林独奏屋，都验证了这一点。在这两个项目中，圆形的混凝土板悬停于开放空间之上。在FIM新总部项目中，采用镂空设计的混凝土板使得日光射入室内，并通过一段如雕塑一般的楼梯连接了建筑的两个楼层。在森林独奏屋项目中，混凝土板变成了一个平台，在平台上陈列着雕塑品般的建筑基础设施组件。此外，在这两个设计案例中，建筑师都通过纤细的金属构件实现了水平混凝土构件和地面之间的接触，而这些金属构件强调了混凝土屋顶呈现出的矛盾的失重效果。在这两个案例中，平面布局巧妙地利用曲线，永远在提醒着人们建筑想要超越玻璃墙界定的物理界限的愿望。

3. 无声的平面

在上面讨论的两节（倾斜功能和曲线）中，建筑师有意突破传统的混凝土墙体、楼面和屋顶的设计方法。因此，他们充分探索了混凝土的建筑表现潜力。然而，也有某些设计案例以更加传统的方式探索了混凝土的视觉和触觉品质。久保田建筑工作室设计的FU住宅位于日本的周南市。在该项目中，混凝土板与定义了住宅外壳的L形白色构件交织在一起。建筑师旨在探索白色表面和留有模板标记纹理的混凝土板

Museum, for example – the use of concrete in Chapel Psalm 36:7 enables the blending of vertical and horizontal elements. Floors, walls and roofs become part of a continuous element that challenges gravity and dissolves the boundaries between the built mass and the surrounding landscape.

The curves in the two projects discussed above are associated with a complex tridimensional geometry. Consequently, they generate powerful formal statements. However, the potential to produce expressive volumes and spaces can also be accomplished by the most basic two-dimensional curvilinear shape, the circle. The new headquarters for the FIM (Federation Internationale de Motocyclisme) designed by the Swiss office Localarchitecture, and the Solo House designed by the Belgian office KGDVS testify to this. In both projects circular concrete slabs hover over an open space. In the new FIM headquarters, these slabs are carefully pierced to let the light come inside and to articulate the two levels of the building through a sculptural stair. In the Solo House, the slab becomes a platform where the infrastructural components of the building are displayed as sculptures. Further, in both cases the contact between the horizontal concrete element and the ground is made through slim metallic elements that emphasize the paradoxical weightlessness of the concrete roof. In either case the layout of the plan is cleverly made to take advantage of the curve as a permanent reminder of the building's willingness to expand beyond the physical limits defined by the glass walls.

3. Silent Plans

In the two sections discussed above – the oblique function and the curve – the concrete was deliberately used to go beyond traditional ways of designing walls, floors and roofs. The concrete's expressive potential was thus fully explored. There

1. Adrian Forty, *Concrete and Culture: A Material History* (Reaktion Books), p.79~80, 2012
2. Paul Virilio, *"The Oblique Function" in Architecture Culture*, ed. Joan Ockman (New York: Rizzoli, 1993), 410–11., 1943-1968
3. Oscar Neimeyer, *The Curves of Time: The Memoirs of Oscar Niemeyer*, Phaidon Press, 2007
4. Forty, *Concrete and Culture*, 279

之间一种细腻的对比。这两种构件形成了与周围环境之间的一种无声的捉迷藏游戏。

一场混凝土的复兴

20世纪70年代，现代主义的整体逻辑受到了挑战，混凝土的使用与一种日渐衰败的建筑设计方法联系在了一起。自20世纪90年代初以来，艾德里安·福尔蒂主张："在世界上这些混凝土早就不再受欢迎的地方，已引发了一场应用露石混凝土的复兴运动。"[4]

本章介绍的项目事实上证明了在当代建筑实践过程中，露石混凝土越来越受欢迎。上面讨论的项目尤其体现了建筑师如何利用混凝土的极端可塑性来创造复杂的几何构造，或仅仅简单地挑战材料的构造。原本沉重的建筑呈现出悬停于开放景观上方的一个轻盈圆盘的效果；而一栋看起来巨大的建筑，则呈现出一种光和影相互作用的多孔箱形结构。这些只是验证钢筋混凝土作为结构媒介的多功能性以及作为文化媒介的柔韧性（就像艾德里安·福尔蒂所说的那样）的一些设计案例。我还要补充一点，它还提供了混凝土作为建筑表现媒介的潜力的有力证据。如果康问道："你想要什么，混凝土？"混凝土可能会说："我想要创造空间。"

are, however, cases in which the concrete's visual and tactile qualities can be explored in more conventional approaches. In the FU-House designed by Kubota Architect Atelier for the Japanese city of Shunan, concrete slabs are intertwined with the L-shaped white elements that define the house's shell. The architects explore a smooth contrast between the white-washed surfaces and the concrete slabs textured with the formwork marks. These two elements produce a silent game of hide and seek with the surrounding context.

A Concrete Renaissance

In the 1970s, while the whole logic of modernism was being challenged, the use of concrete became associated with a decadent architectural approach. Since the early 1990s, Adrian Forty argued, "In those parts of the world where the concrete had earlier gone out of favour, there has been a renaissance in the use of exposed concrete."[4]

The projects featured in this section testify indeed to the increasing popularity of the exposed concrete in contemporary architectural practices. In particular, the projects discussed above show how architects take advantage of the concrete's extreme plasticity to produce complex geometric compositions, or to just simply challenge the tectonics of the material. What is heavy comes about as a light disc hovering over an open landscape; What looks like a monolithic block reveals itself as a porous box of light and shadows. These are just some examples that testify to the reinforced concrete's versatility as a structural medium and its pliability as a cultural medium, as Adrian Forty put it. It delivers also, I would add, strong evidences of the concrete's potential to perform as a medium for architectural expression. If Kahn would have asked "What do you want, concrete?" Probably the concrete would say: "I'd like space." Nelson Mota

萨维耶斯某住宅

Anako'architecture

这所新建的家庭住宅坐落于一个用途并不明确的场地。

尽管周边的独立式别墅使场地的俯瞰视野受到了限制，但是它位于南向的缓坡之上，在这里可以出人意料地纵览阿尔卑斯山脉以及罗纳河左岸的美景。

项目的设计符合场地的实际条件，并肯定了它的激进特色，其目的旨在通过捕捉和提升内在品质的方式来摆脱场地的限制。

场地的边界采用无窗的清水混凝土墙，它界定了住宅的立面，好似城堡一般，未设置任何开口，或者说是不与周围的房屋产生任何直接的联系。在靠近场地北部的上坡区域，中空的楼盖定义了路边的建筑主入口。该住宅体现为一个单层的单体建筑体量，依偎于平缓下降的地面之上，将内部连通、被天井隔开的三个半层内部体量连成一条直线。这些半层分别设置有入口、夜间和白天活动区。

项目遵从了通过认识和尊重场地规模而实现一体化设计的设计委托。不过，它拒绝模仿周边的任何建筑。

然而，矛盾的是，建筑物厚重的外表与提供通向外部世界的宽敞通道的内部开口形成了鲜明的对比。在混凝土外壳的中心设计的三个天井使日光可以洒满纯净的白色内部空间。这些空间起到了室内花园的作用，将绿茵和蓝天引入住宅中央，在室内外空间之间营造了一种透明的感觉。

天井的混凝土护墙位于每一个半层之间，恰好勾勒出天边山脉的轮廓。三个天井将整体的砾石屋顶分开，而屋顶作为第五立面顺应并强化了建筑体量带有开口的自由几何结构。

清水混凝土墙通过采用古旧的、遭到严重破损的木质表面衬层的金属模板来定形。如此一来，在混凝土的表面形成一种神奇而独特的浮雕装饰纹理，从而使每一次的模板拆除都成为产生新发现的时刻。

各种尺寸模板的随机组装，加之混凝土的微小颤动露出了砂砾底层，给表面带来了任意风格的雕塑装饰图案，随着时间和季节的变化而呈现出多姿多彩的面貌。在该项目中，混凝土既是建筑的结构，也是材料；既塑造了空间，也塑造了光影。它将建筑概念的激进特色具体地表现了出来。

House in Savièse

The new family home is set on a site characterized by a relative ambiguity.

Although the plot suffers from a tight overlook with the neighboring detached villas, it also benefits from south-facing orientation on a gentle slope, which offers an unexpected overview on the alpine mountains and along the left bank of the Rhône River.

The project consents to its condition and affirms its radicalness seeking to free itself from the limitations of the site, by capturing and enhancing its intrinsic qualities.

Blind walls of raw concrete border the site and define the façade of the house which, like a fortress protects itself from any opening to, or direct relation with, neighboring properties. To the north, uphill of the plot, a hollow cover defines the main entrance next to the access road. The house appears as a monolithic volume that embraces the gently dipping

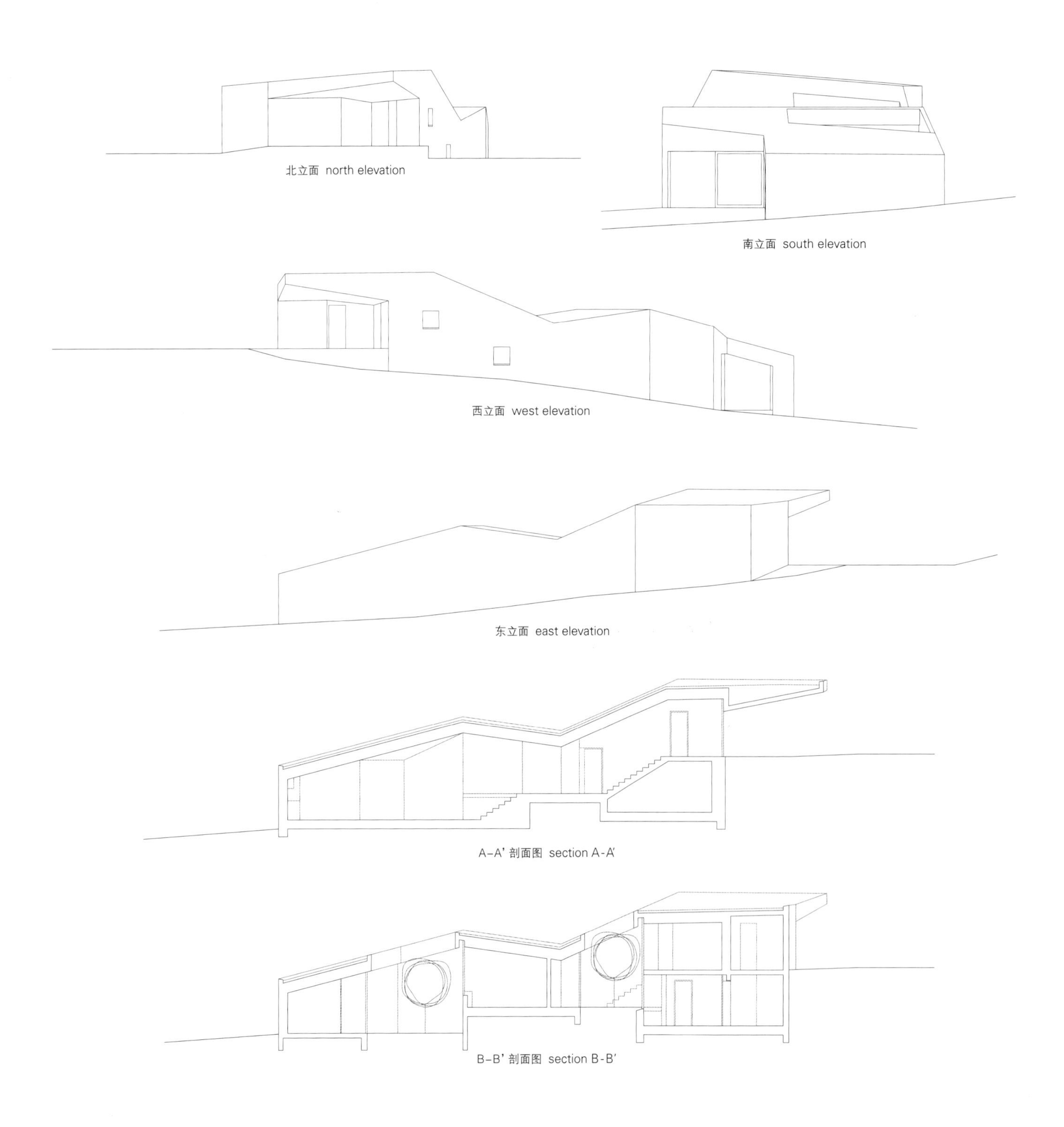

LOUISE

ground on a single level, articulating internal volume as an alignment of three interconnected half-levels separated by patios. These half-levels define the entry, night and day areas. The project complies with its commitment to integration by recognizing and respecting the scale of the site. It rejects, however, any mimicry with neighboring buildings.
Paradoxically, the massive external appearance of the object contrasts with inner apertures that offer generous access to the external world. In the core of this concrete shell, the three patios allow daylight to flood white and pure internal spaces. These spaces act as interior gardens that invite the greenery and blue sky into the heart of the house, creating a feeling of transparency between internal and external spaces.
Concrete parapets of the patios are located between each half-level and precisely frame the outline of mountains in the sky. A gravel roof is dissected by the 3 patios and stands as a 5th facade that adopts and reinforces a free, cut-out geometry of the volume.
Walls of raw concrete were shaped by metallic formwork with worn and heavily damaged wooden skin lining. This gives unexpected and unique reliefs and textures on the surface of the concrete; each formwork removal was a genuine moment of discovery.
Random assembly of formwork of various sizes combined with minimal vibration of the concrete revealing gravel nests, give the surface random reliefs and patterns that express multiple colours and shades through the day and seasons.
Concrete is structure, material, space, shadow and light. It materializes the radicalness of the architectural concept.

项目名称：House in Savièse / 地点：Savièse, Switzerland / 建筑师：Anako'architecture / 结构工程师：SD ingénierie SA / 用地面积：656m² / 总建筑面积：244m²
外部装饰：exposed concrete / 室内装饰：plaster and concrete / 施工时间：2015.8—2016.9 / 摄影师：©Nicolas Sedlatchek(courtesy of the architect)

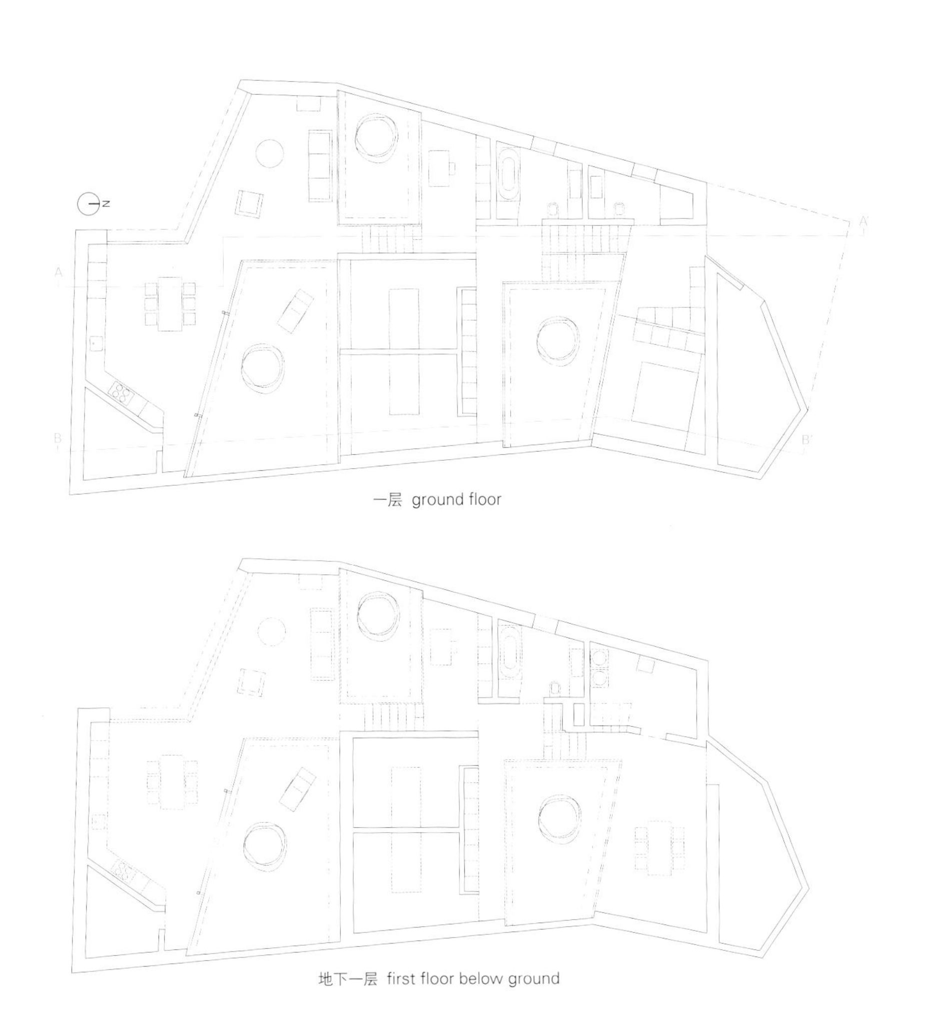

一层 ground floor

地下一层 first floor below ground

生态观测住宅

archipelagos - Manolo Ufer

生态观测住宅被构思成一种复杂的建筑界面，它经过优化可引导能量流动并收获现场的物力资源。本项目的名称是一个复合词，取"生态"和"视野"的词源含义。"生态"源自希腊语"oikos"，泛指与其环境相关的生物。而"视野"意为相关标的物可以在其内进行操作或建立框架的区域范围。这两个术语合在一起创造了一种理念，它并没有将建筑解读为一个静态物体，而是一个媒介或界面，人们通过它可以（重新）定义与周围环境的关系。

流动的空间

该住宅位于墨西哥蒙特雷的南部郊区，在马德雷山脉的脚下。这种接壤的情况要求建筑师采取一种新的设计方法来处理场地与其周围环境的关系。

在这些前提下，就要求对其进行超越地块法定条件的重新诠释。然而，此种扩充的定义并不是指将周边环境简单地与设计相结合，而是被解读为一个由逻辑和当地物流（利用多种方法来管理多变的资源流动）定义的抽象型多维建筑空间，即一个生态系统。

本土建筑

在自然与人工、城市与山脉的交界处，在捕捉不同生态系统的变化之时，该住宅成为一个互换平台的集合体。在当今全球信息时代的背景下，项目重新考虑了场所性和文脉主义的问题。总之，该住宅旨在实现具有"墨西哥"元素的建筑。

顺着这条逻辑思路，该项目以当地条件为基础，成为一栋本土化建筑。这些空间对周围环境中现有的资源和能量做出了特殊回应，因此成为一个本地化元素的集合体。

形态形成

这栋住宅的最终效果也是多个输入变量的形态形成的过程。在该项目中，建筑师分析了几何体的日光照射入口和热辐射，检测了盛行风、当地的山谷气流和当地山风、地表水流模型和雨水沟径流，并且考虑了逐渐增多的人类活动的影响。经过这些复杂的评估，建筑师将承重面与不同厚度的交错楼板、横梁综合起来，以确保实现相互稳定性。最终形成了一个带有连续钢筋混凝土外壳的建筑集合体。

规划

这栋住宅位于一块占地1251m²的场地之上，总建筑面积651m²。外部空间占据首层超过一半的面积。内部是一个宽敞的开放空间，用作居住区，并带有一个门廊，可以遮阳挡雨，还可以纵览Huajuco峡谷的风景。服务区包括配备食品储藏室的厨房以及洗衣房。住宅的该部分包含一个可停放两辆车的屋顶车库，这个车库又挨着另一处户外停车场，还能再停放两辆车。在住宅的上层，可用室内面积达199m²，这里由三个卧室套间组成，其中包括一间主卧，带有能够欣赏壮观山脉景观的种植阳台。

混凝土

从结构的角度到相关的日光照射量考虑，建筑师对每一面墙的厚度都有明确规定。如此一来，该建筑汇聚了众多厚度为110～350mm的墙体。

建筑的形态向传统的钢筋混凝土施工方式提出了挑战。建筑师根据相关平面和建筑构件的结构设计说明，采用了具有不同强度的各种

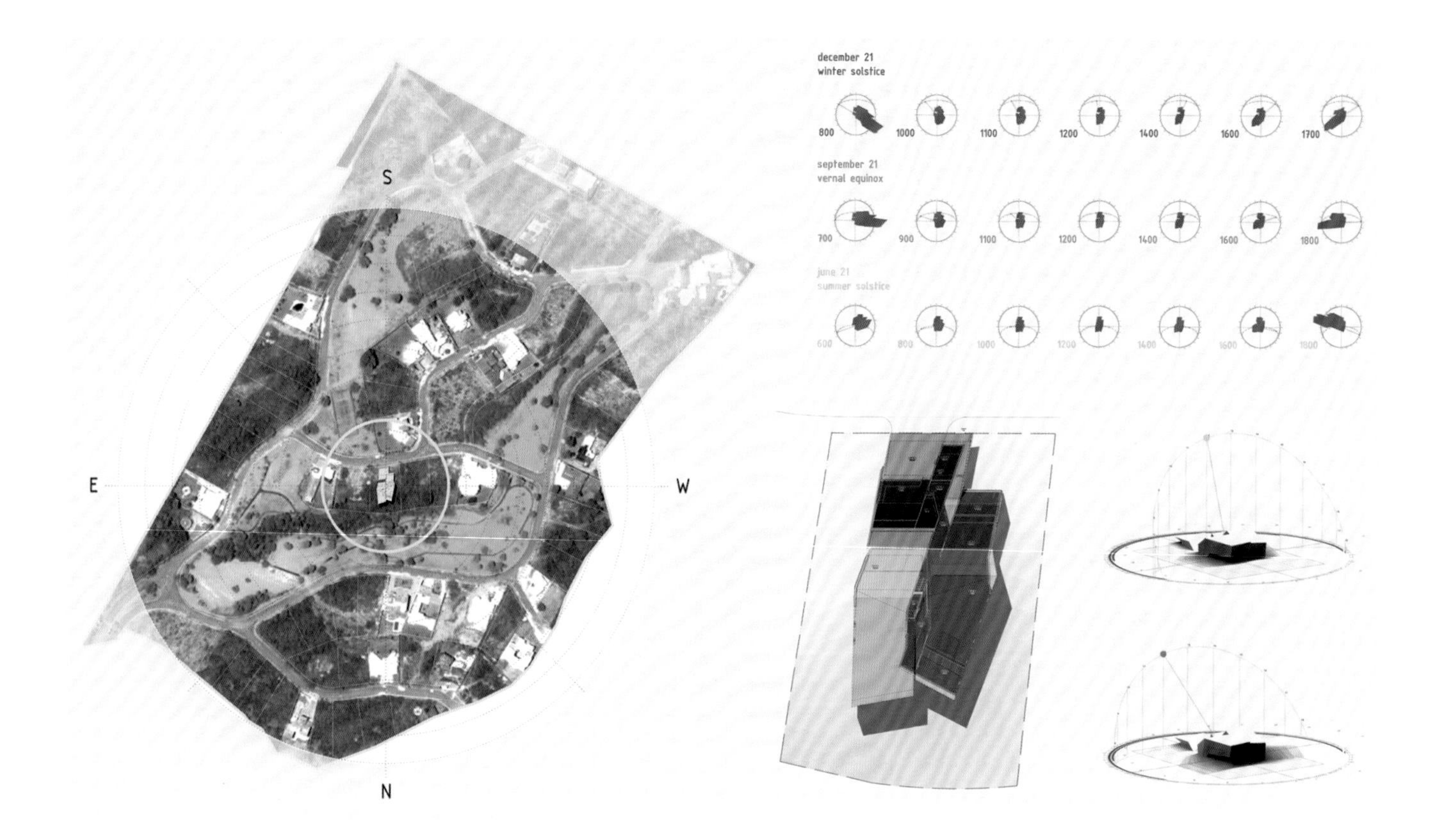

形式分析
formal analysis

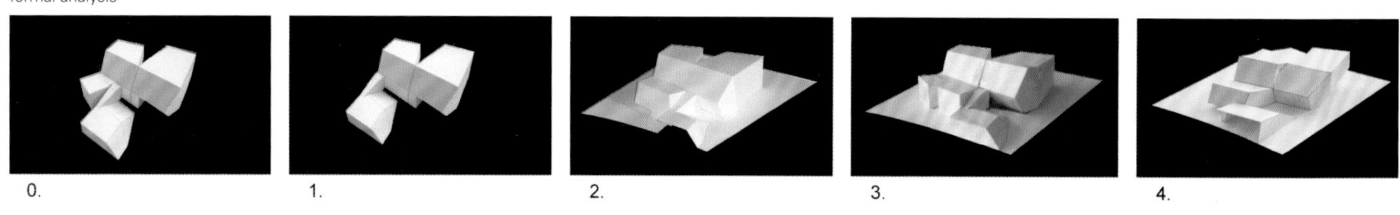

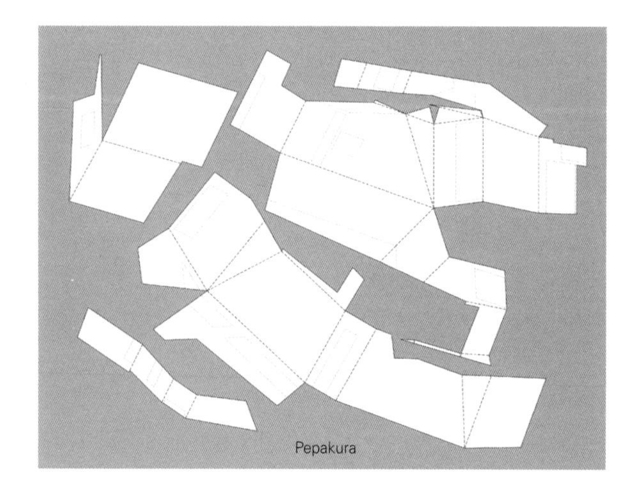

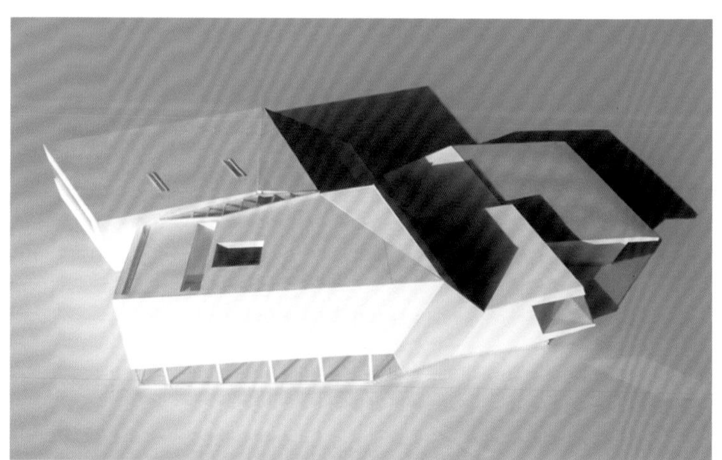

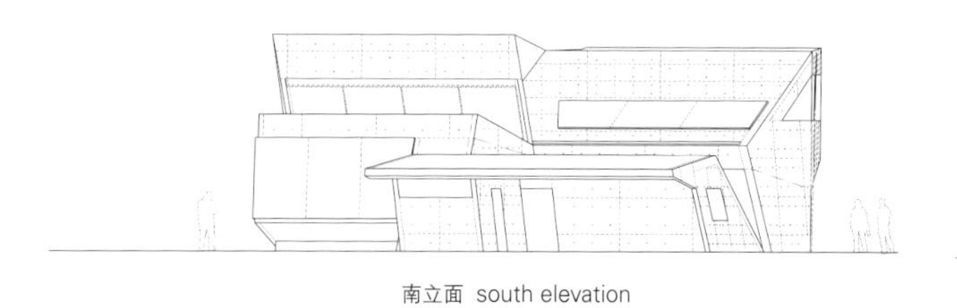

南立面 south elevation

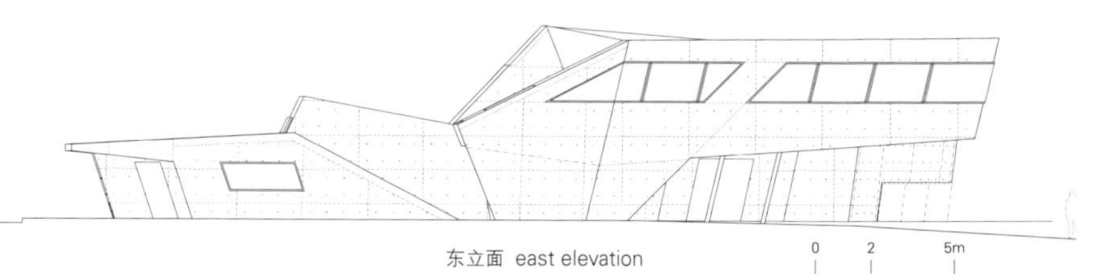

东立面 east elevation

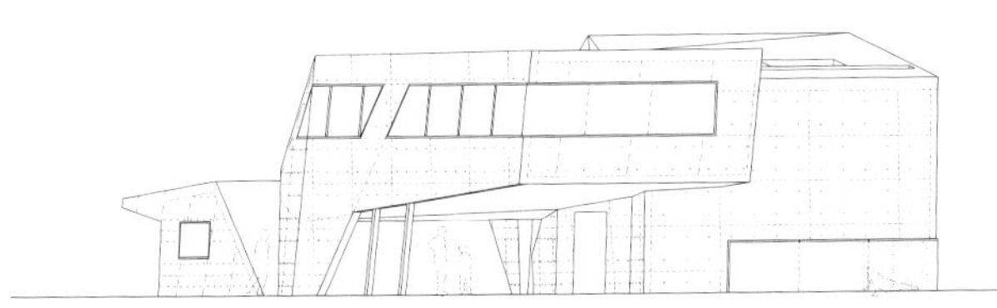

北立面 north elevation

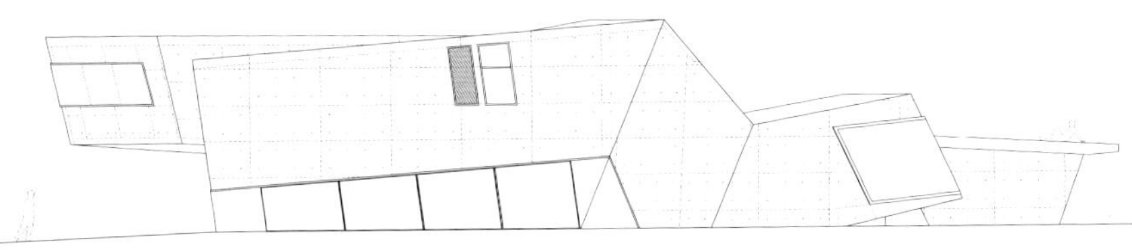

西立面 west elevation

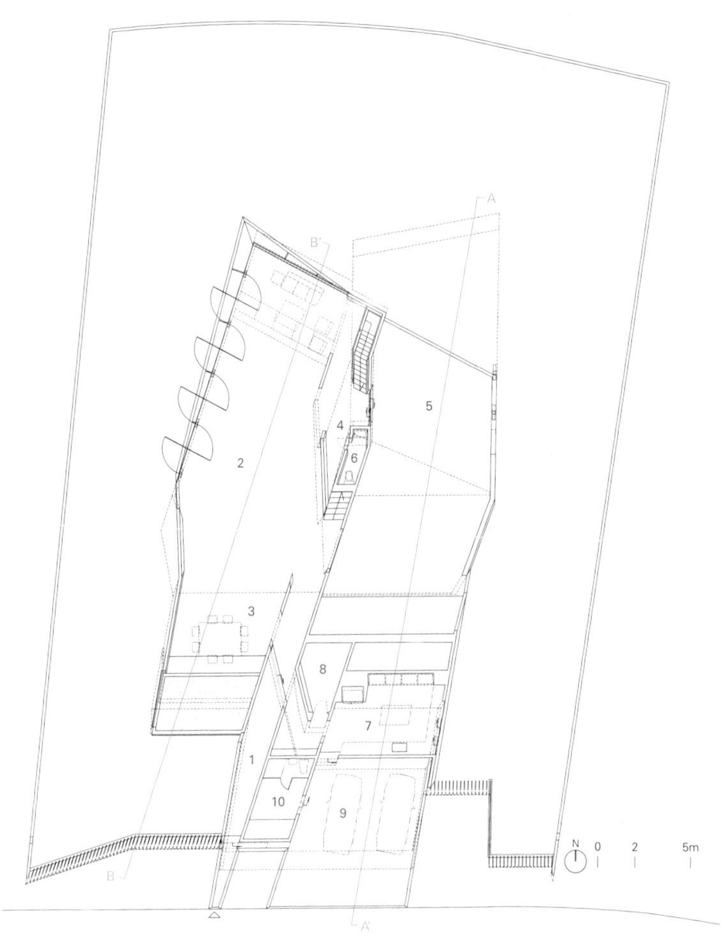

1. 入口走廊 2. 起居室 3. 餐厅 4. 中庭 5. 门廊 6. 卫生间 7. 厨房 8. 食品储藏室 9. 简易车库 10. 设备室
1. access gallery 2. living room 3. dining room 4. atrium 5. porch 6. toilet 7. kitchen 8. pantry 9. car port 10. service room
一层 first floor

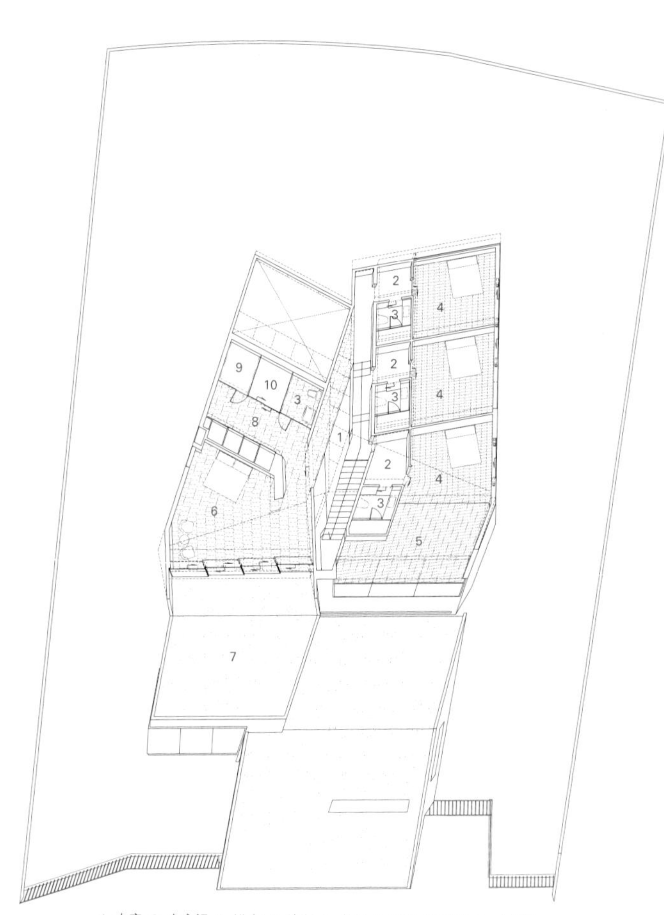

1. 中庭 2. 小房间 3. 浴室 4. 卧室 5. 书房 6. 主卧 7. 露台 8. 小房间 9. 冲凉房 10. 户外冲凉室
1. atrium 2. closet 3. bathroom 4. bedroom 5. study 6. master bedroom 7. terrace 8. closet 9. shower room 10. outdoor shower
二层 second floor

混凝土。而在应用更多的传统工艺之前，建筑师使用了一个虚拟模型来转变和控制建筑的几何结构。

实践和环境

该住宅使用了钢材、玻璃和混凝土这样的基本材料，并且避免使用现成的解决方案，或考究的材料以及艳丽的饰面。

然而，因为结构的复杂性以及采用了现浇钢筋混凝土结构，该项目的成本和预算都所有增加。而得益于总体积为445m³的混凝土产生的被动热控制和自然通风，建筑减少了对主动供暖和空气调节系统的依赖，而这其中还不包括对水资源的再利用。

Ecoscopic House

The Ecoscopic House is conceived as a complex architectural interface optimized to channel flows of energy and harvest material resources of the site. The name is a compound from the etymological meanings of "eco-" and "-scope". "Eco-", from the Greek word "oikos" for household, concerns with living things in relation to their environment. While "-scope" is the extent of the area on which the relevant subject matter operates or is framed within. These two terms come together to create a concept that understands architecture not as a static object but a medium or interface through which man (re)

defines the relationship with his surroundings.

Space of Flows

The house is located in the southern suburb of Monterrey, Mexico, at the feet of the Sierra Madre mountain range. This bordering condition demanded a new approach towards the relation between the site and its surrounding environment. With these premises, a reinterpretation that extends beyond the legal terms of the plot was called upon. However, this expanded definition is not a mere gesture in incorporating the immediate context into design. Instead, it is understood to be an abstract multi-dimensional architectural space defined by the logic and logistics of a territory that manages variable flows of resources at multiple scales; an ecosystem.

Indigenous Architecture

At the threshold between the natural and artificial, the city and the sierra, the house becomes an assemblage of interchanging platforms, as it tries to capture the flux of different ecosystems. The project re-evaluates questions of locality and contextualism within today's global information age. In all, the house asks what could be considered "Mexican".

Along this line of logic, the project becomes indigenous to the site as it roots and orientates itself on local conditions. The spaces specifically respond to available resources and energies within the surrounding environment, thus becoming a vernacular spacecraft.

Morphogenesis

The house is also a result from a morphogenesis process of multiple input variables. Analysis of geometric solar access and thermal radiance, examination of prevailing winds, regional valley currents and local mountain breezes, model of surface water flow and storm drain runoff and the impact of continuously encroaching human activity were considered in the project. Through this complex assessment, loadbearing planes along with interlacing slabs and beams of variable depth were synthesized together to ensure mutual stability. The result was an assemblage conceived as a continuous reinforced concrete enclosure.

Program

The house stands on a 1,251m² plot with a total built area of 651m². The ground floor allots more than half of its area to the exterior space. The interior is a vast open space that serves as a living area complemented by a porch protected from the sun and rain with panoramic views of the Huajuco Canyon. The service sector integrates the kitchen with the pantry and laundry room. This part of the house is coupled with a two-vehicle roofed parking area adjacent to an additional open-air parking for two more cars. The upper floor, amounting to 199m² of usable interior space, incorporates 3 ensuite bedrooms and includes a master bedroom which boasts a planted terrace with spectacular views of the sierra.

Concrete

The thickness of each wall is specified from both standpoint of structure and relevant exposure to solar radiation. In this way, the building draws together a variety of walls with thicknesses between 110mm to 350mm.

The form of the building challenges the traditional methods of reinforced concrete construction. Different concrete strengths were employed in accordance to structural specifications of relevant planes and building members. A virtual model was used to translate and control the geometry of the building before more traditional technologies were applied.

Realization and the Environment

The house uses elementary materials such as steel, glass, and concrete. The project avoids off-the-shelf catalogue solutions, or sophisticated materials and voluptuous finishes.

However, due to the structural complexity and in-situ reinforced concrete constructions, there were increases in costs and budgets. The total of 445m³ of concrete is an investment paid off by passive thermal control and natural ventilation that decreases dependency on active heat and air conditioning systems, while not excluding water reuse.

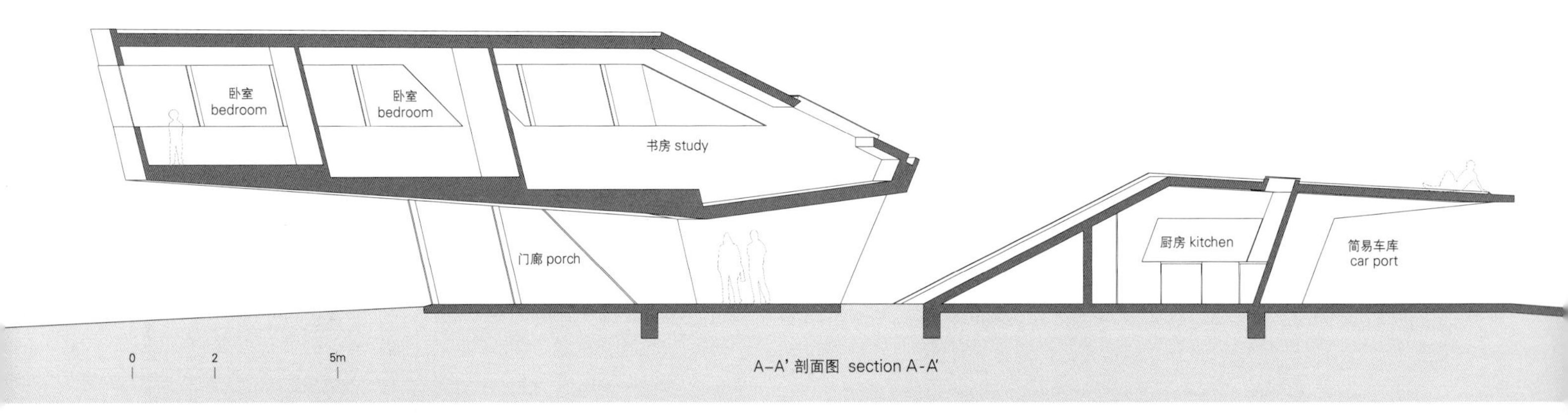

A–A' 剖面图 section A-A'

项目名称：Ecoscopic House / 地点：Monterrey, Mexico / 建筑师：archipelagos
项目团队：Manolo F. Ufer, Alberto Sanchez, Angela Soong / 合作方：Vivie Lin, Enrique Moya-Angeler, Juan Angel Castañeda, Raul Casillas, Claudia María López Beltrán, Eduardo Peña, Rodrigo Tello Medina, Joaquín Legorreta, Ariadne Flores, Jose Maria Lozano, Benjamin Dávila, Carolina Ayala, Arturo Astiazarán, Guillermo Ortiz, Elsa Mendoza, Daniel Pistoff, Roberto García, María Paz Argüello, Laura Rojas, Fernando Granados, Rafael Miranda Acuña, Alberto Rodriguez, Maximino Zapata Ramos, Humberto Tamez Flores, Rafael de la Garza, Anuar Rios, Ciro Alfano / 照明设计：VV Lin / 承包商：Simon Niño / 模型：Saturnino Ipiña
家具：Roche Bobois / 供应商：CEMEX, concrete; Ternium, steel; Vitro, glass; Mármoles Regiomontanos, stone; Owens Corning, insulation; BASF Master Builder Solutions, waterproofing; Cuprum, aluminum windows; Hafele, door hardware; Blum, kraus, kitchen fittings; Bosch, general electric, kitchen appliances; Hansgrohe, Roca, bathroom fittings; Lutron, lighting control; Phillips, light fixtures; Roche-Bobois, furniture / 用地面积：1,251m² / 建筑面积：651m² / 设计时间：2008 / 施工时间：2008—2016
摄影师：©Roland Halbe

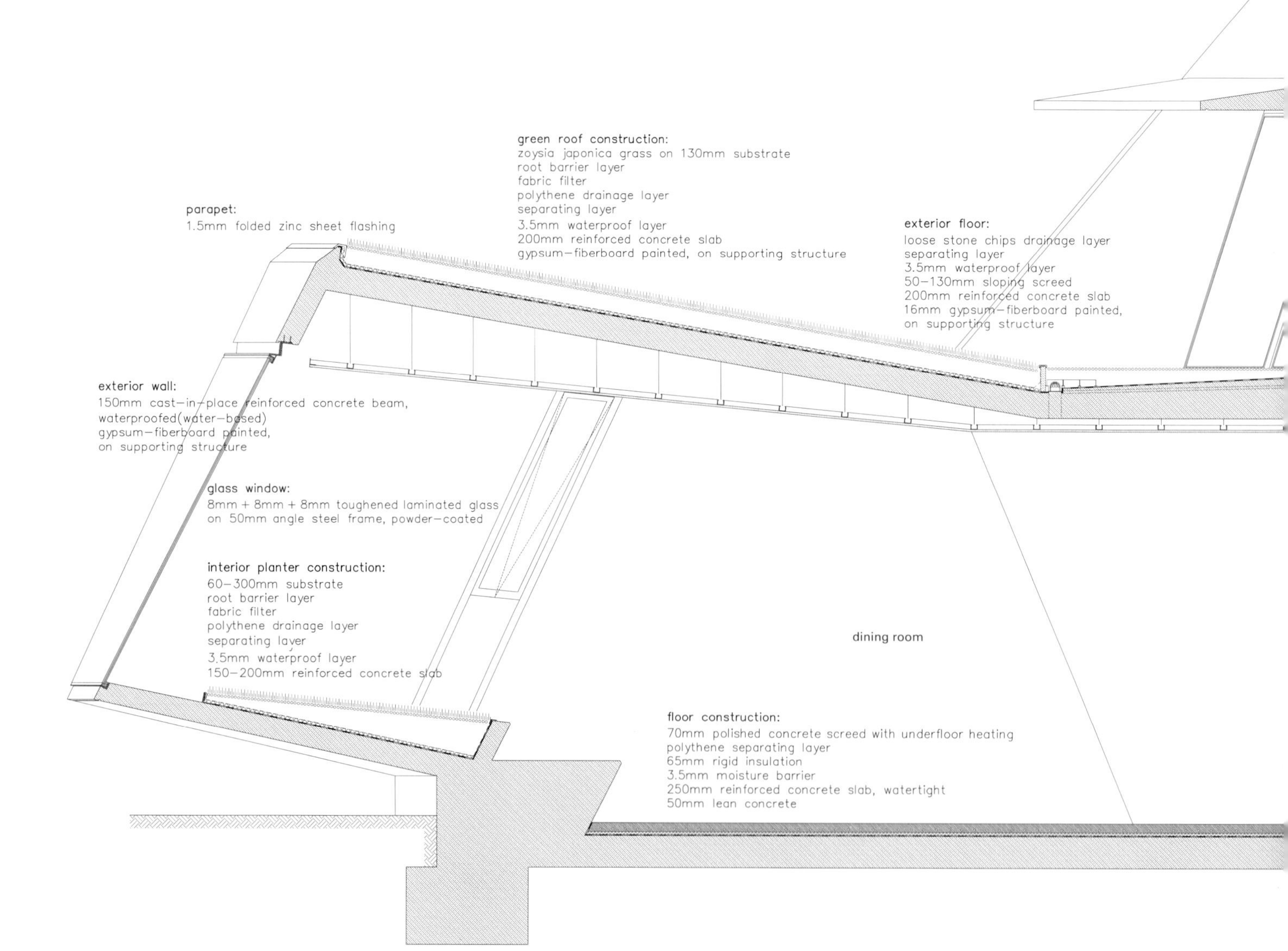
green roof construction:
zoysia japonica grass on 130mm substrate
root barrier layer
fabric filter
polythene drainage layer
separating layer
3.5mm waterproof layer
200mm reinforced concrete slab
gypsum-fiberboard painted, on supporting structure
parapet:
1.5mm folded zinc sheet flashing
exterior floor:
loose stone chips drainage layer
separating layer
3.5mm waterproof layer
50-130mm sloping screed
200mm reinforced concrete slab
16mm gypsum-fiberboard painted,
on supporting structure
exterior wall:
150mm cast-in-place reinforced concrete beam,
waterproofed(water-based)
gypsum-fiberboard painted,
on supporting structure
glass window:
8mm + 8mm + 8mm toughened laminated glass
on 50mm angle steel frame, powder-coated
interior planter construction:
60-300mm substrate
root barrier layer
fabric filter
polythene drainage layer
separating layer
3.5mm waterproof layer
150-200mm reinforced concrete slab
dining room
floor construction:
70mm polished concrete screed with underfloor heating
polythene separating layer
65mm rigid insulation
3.5mm moisture barrier
250mm reinforced concrete slab, watertight
50mm lean concrete

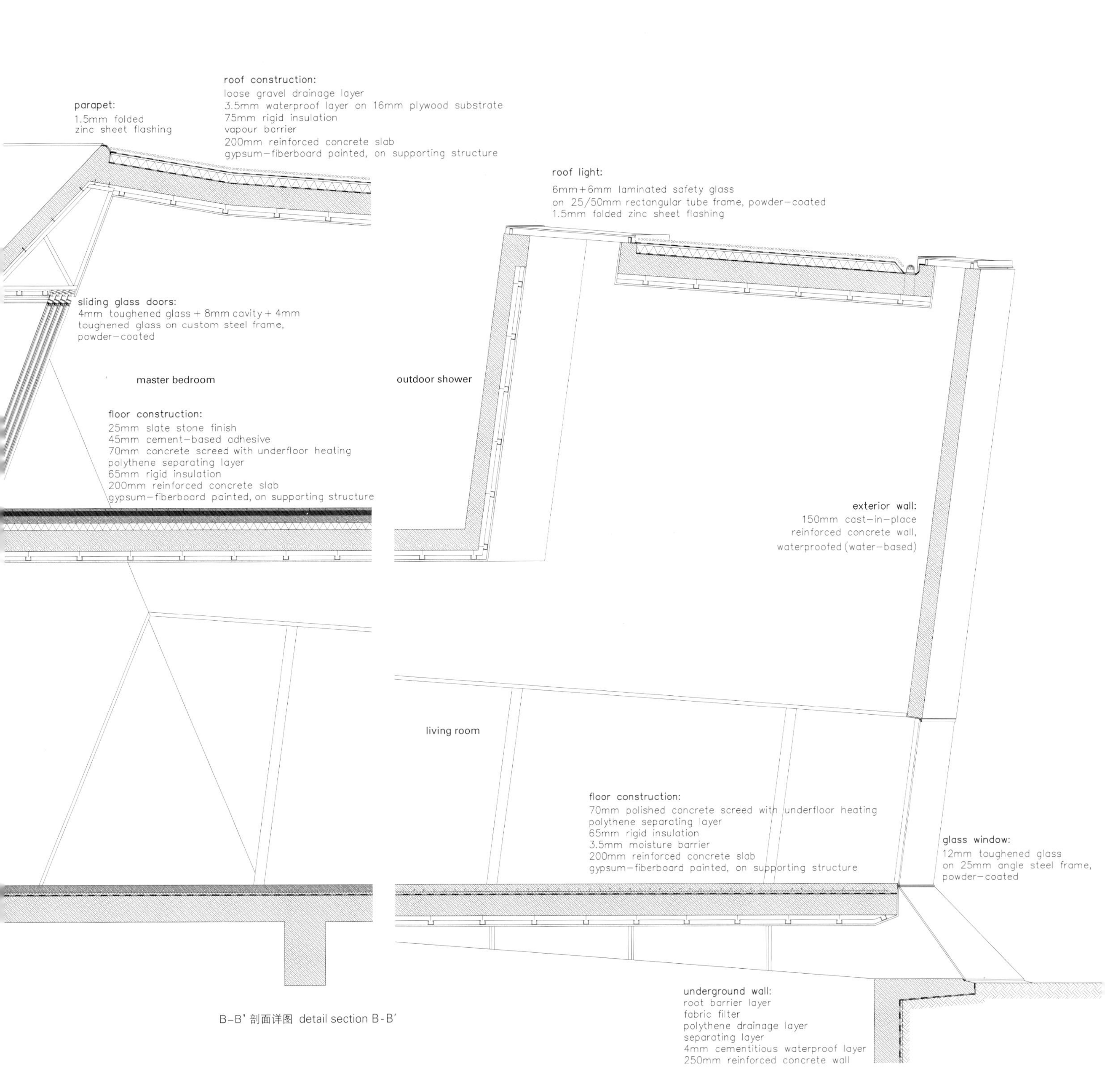

B–B' 剖面详图 detail section B-B'

U Retreat 度假屋

IDMM Architects

洪川郡垈谷里，其字面意思为"阔湖"和"高丘"，暗指这块土地及其地理性质。Sari谷是Sori山脉一处海拔100m的峭壁，从那里可以俯瞰40m之外的项目场地。巨石嶙峋的峭壁的规模以及不知不觉变得繁茂的植被的生命力，更使得周围的环境充满了生机。

当风吹过山谷的时候，整个悬崖仿佛都随着撼动。树木在风中摇曳，根据其重量摆动的幅度各有不同，这些微微摇曳的姿态共同形成了悬崖上一股流动的波浪。如何通过建筑设计来诠释和体现气候，以及随着每一瞬间变化的悬崖的运动和波浪？

该项目是在一位从时尚产业退休的商人的委托下建造的。退休后，他想要一座位于山谷深处、独特而又适合冥想的住宅。项目的重点在于如何将一种小型的当地住宅，也就是所谓的"养老屋"转变成一处特别的处所。建筑师先是诠释了其对大自然的独特理解，再分析了从住客的位置上所看到的各式风景，以此为住客提供了一种全新的空间体验。

每个人来访度假屋的目的不尽相同，有一家人来此旅行只住一晚，也有情侣来此约会。他们都想在无人打扰的情况下在一个精心设计的空间内休息。住客可以体验一种全新的空间性，在那里，客厅、主卧室、韩式卧室、浴室、露台和水疗设施都位于不同的楼层，而非在同一楼层上用隔墙分开。多楼层设计的原因在于住客在各个楼层上都可以与自然风光之间产生多元化的联系，从多种角度欣赏大自然。

从场地望去，不远处的山脉就好像是飞翔中的固态体块。此外，根据预计，由混凝土体块构成的斜线构造可以与大自然不规则而多变的线条产生对比效果。在这方面，倾斜的立柱和悬臂梁，其上下两部分与一种扭曲的姿态相连，共同产生了一种积极的形状变化，与自然风光协

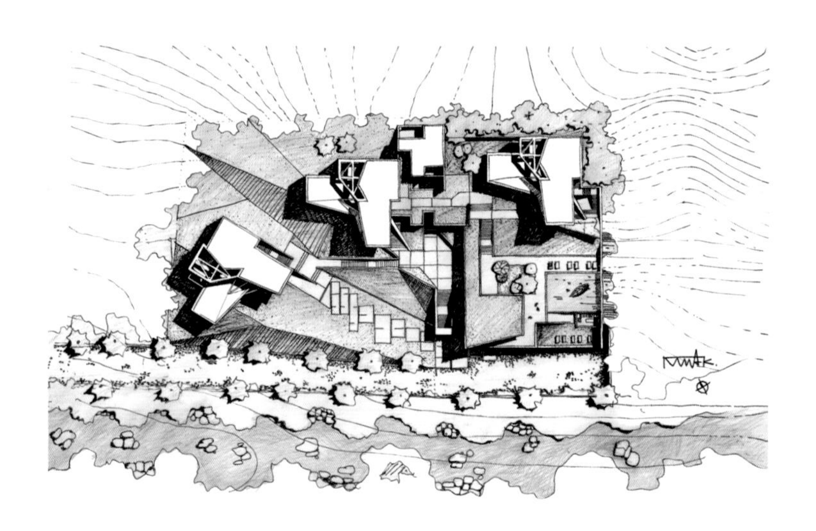

HONGCHEON(洪川) DAEGOK-RI(垈谷), WHICH LITERALLY MEANS "WIDE LAKE" AND "HIGHMOUND", IS A NAME THAT ALLUDES TO THE LONG PERSISTING CONDITIONS OF THE LAND. A BERTICAL CLIFF AT AN ALTITUDE OF 100M, KNOWN AS THE SORI-SAN MOUNTAIN RANGE SARI-GOL VALLEY, LOOKS DOWN ON THE SITE FROM AS LITTLE A DISTANCE AS 40M AWAY. THE SCALE OF THE CLIFF, PINNED DOWN WITH ENORMOUS STONES, AND THE VITALITY OF THE UNWITTINGLY FLOURISHING VEGETATION WITHIN IT OVERWHELMS THE SURROUNDINGS.
U RETREAT

THE CONCEPT OF HEIGHT WAS APPLIED TO EACH OF THE UNITS, SO THAT USERS MAY ENJOY THE SURROUNDINGS AT A VARIETY OF LEVES. EACH SKIPPED UNIT CONCEIVES VARIOUS RETREAT PLACES, DIVERSE LEVELED INTERIOR SPACES, THE SKIPPED TERRACES, A PRIVATE POOL AND SPA AND SO ON.
U RETREAT

调相融。该设计通过尽量减少与表面的接触来应对丘陵地带，它是悬挑的体块与起伏的景观互动的偶然成果。

U Retreat度假屋正对着悬崖。悬崖好似一幅随季节变化的背景图，住客通过U Retreat度假屋的玻璃窗和多层混凝土平台，可轻而易举地将丰富多彩的悬崖景色尽收眼底。

U Retreat度假屋有两种户型：楼上（108m²）和楼下（49m²）户型。尽管在同一个体量之内相互连接，但楼下的户型正对着地面，而且与楼上的户型是隔开的。楼上户型配有空中露台和水疗设施，住客在做水疗时可以冥想、欣赏悬崖和天空的景色。公用设施空间包括门房、咖啡厅以及带小厨房的餐厅。这些设施也支持住客在U Retreat度假屋的范围内开展游泳、烧烤和冥想等户外活动。

U Retreat

Hongcheon-gun Daegok-ri, meaning "wide lake and high hills", alludes to the land and its geography. The Sari-gol Valley, a vertical cliff of Sori-san Mountain with a height of 100m, looks down on the site only 40m away. The scale of the cliff, pinned down with enormous stones, and the vitality of the unwittingly flourishing vegetation within it, overwhelm the surroundings.

THE JENGJA IS A SPACE WHICH IS UNIFIED WITH NATURE, LIBERATING ITS USERS FROM THE CONFINES OF THE HOME. THE PROGRAM OF THE JEONGJA IS REST, PLAY, AND RETREAT WITH THE NATURE. THE ENORMOUS SCREEN ON THE STEEP SITE WAS INSPIRED BY THE CONCEPT OF 'HEIGHT' FORMING A RELATIONSHIP WITH THE CLIFF ALLUDING TO AN INK-AND-WASH PAINTING (GUMI-E).

LI RETREAT IOOWK.

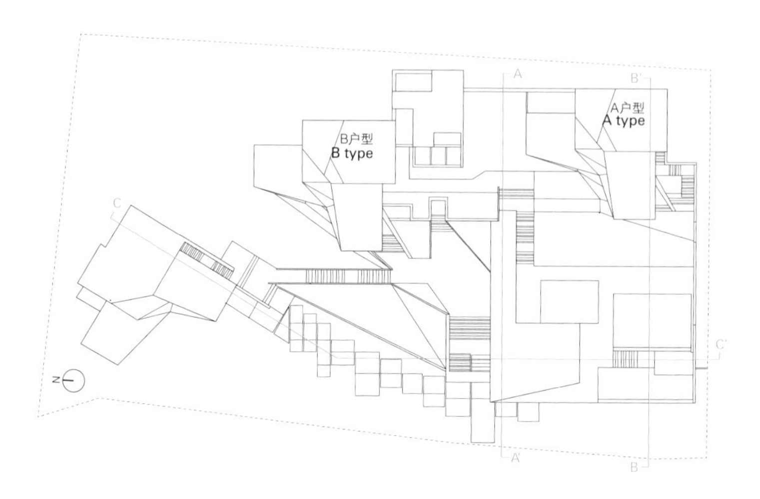

LIKE THE IMAGE OF THE CLIFF AND AN ASSEMBLY OF TREES LI RETREAT IS ALSO COMPOSED OF UNIT ORGANISMS THAT HAVE BEEN TRANSPOSED THROUGH ARCHITECTURE. THE DESIGN OF EACH UNIT WAS INSPIRED BY EACH OF THE MOMENTS THAT SUCH A FLOWING MOTION IS GENERATED.

LI RETREAT IOOWK.

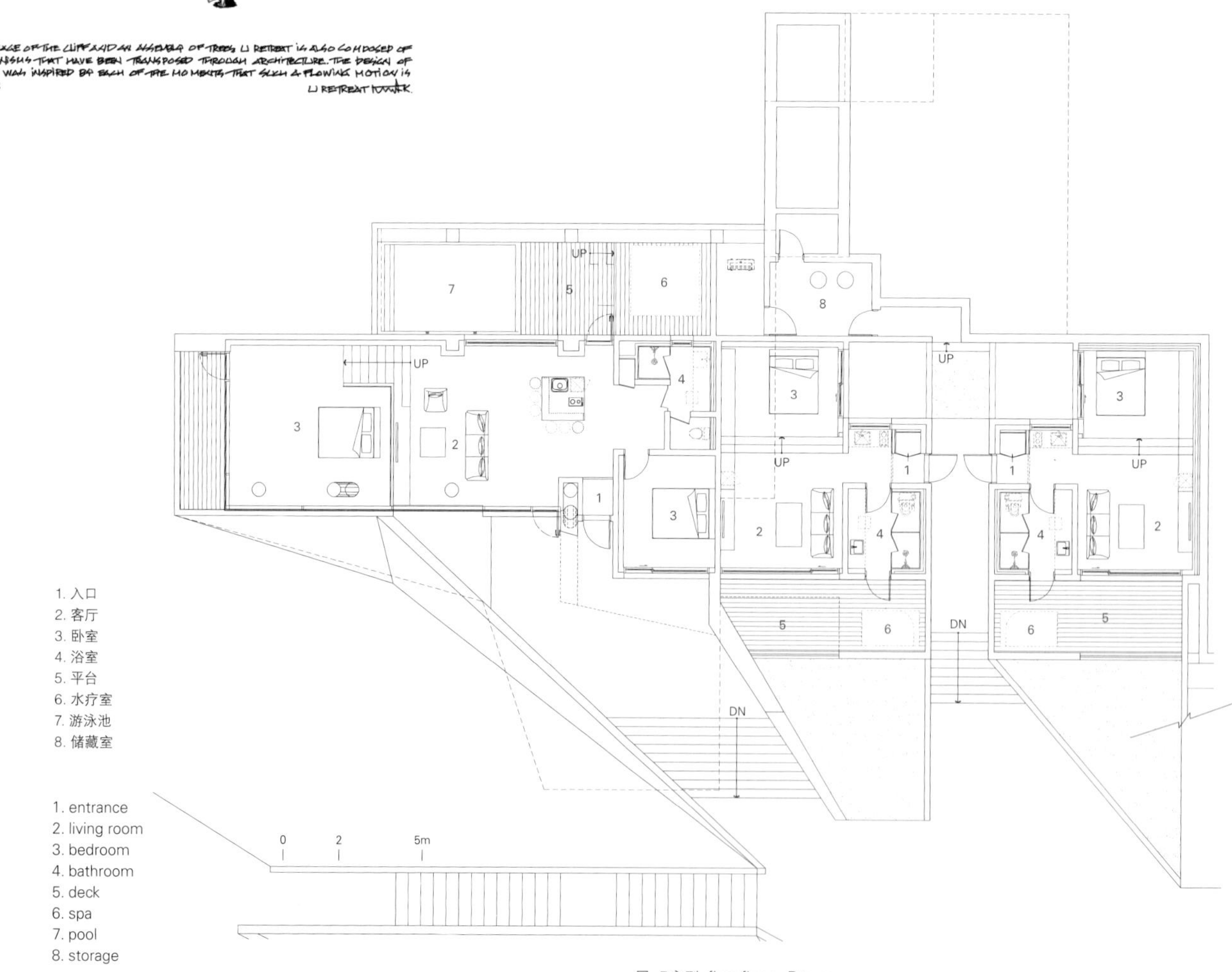

1. 入口
2. 客厅
3. 卧室
4. 浴室
5. 平台
6. 水疗室
7. 游泳池
8. 储藏室

1. entrance
2. living room
3. bedroom
4. bathroom
5. deck
6. spa
7. pool
8. storage

一层_B户型 first floor_B type

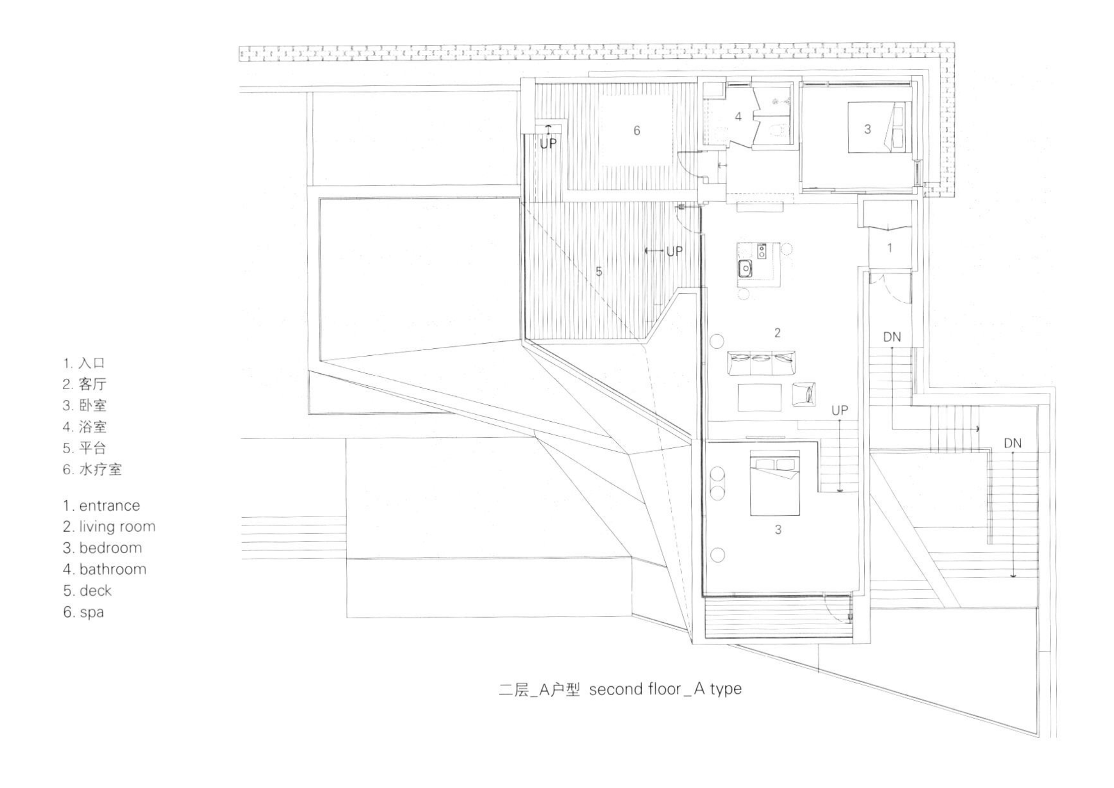

二层_A户型 second floor_A type

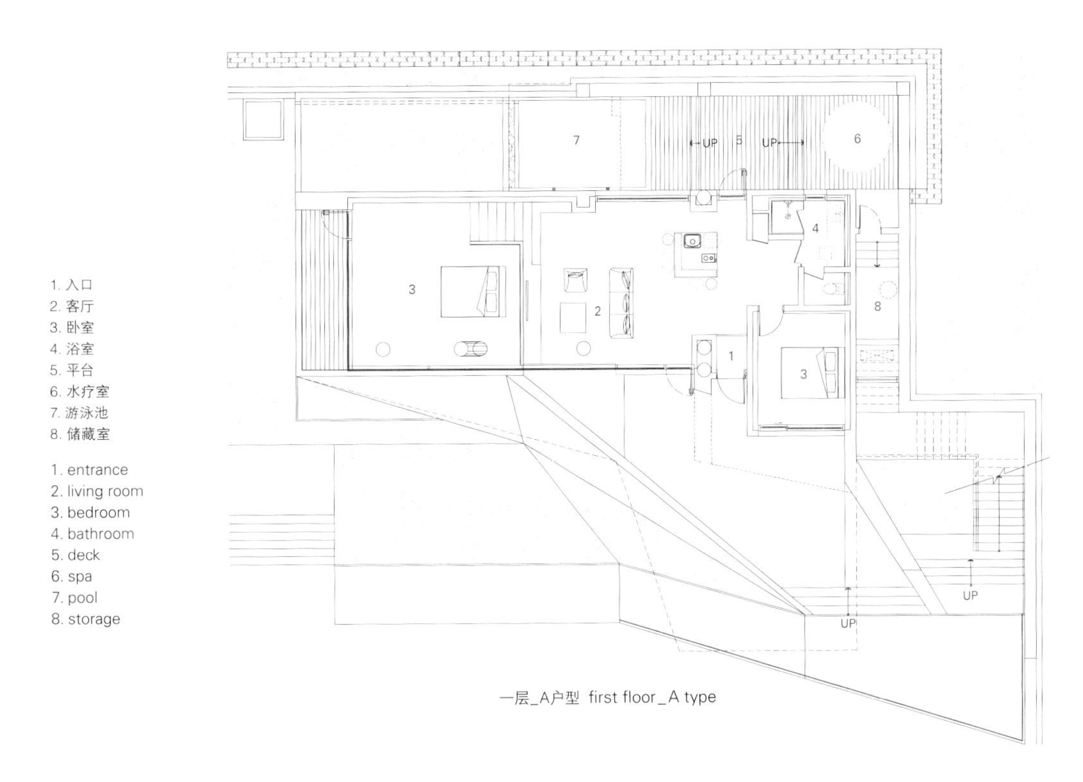

一层_A户型 first floor_A type

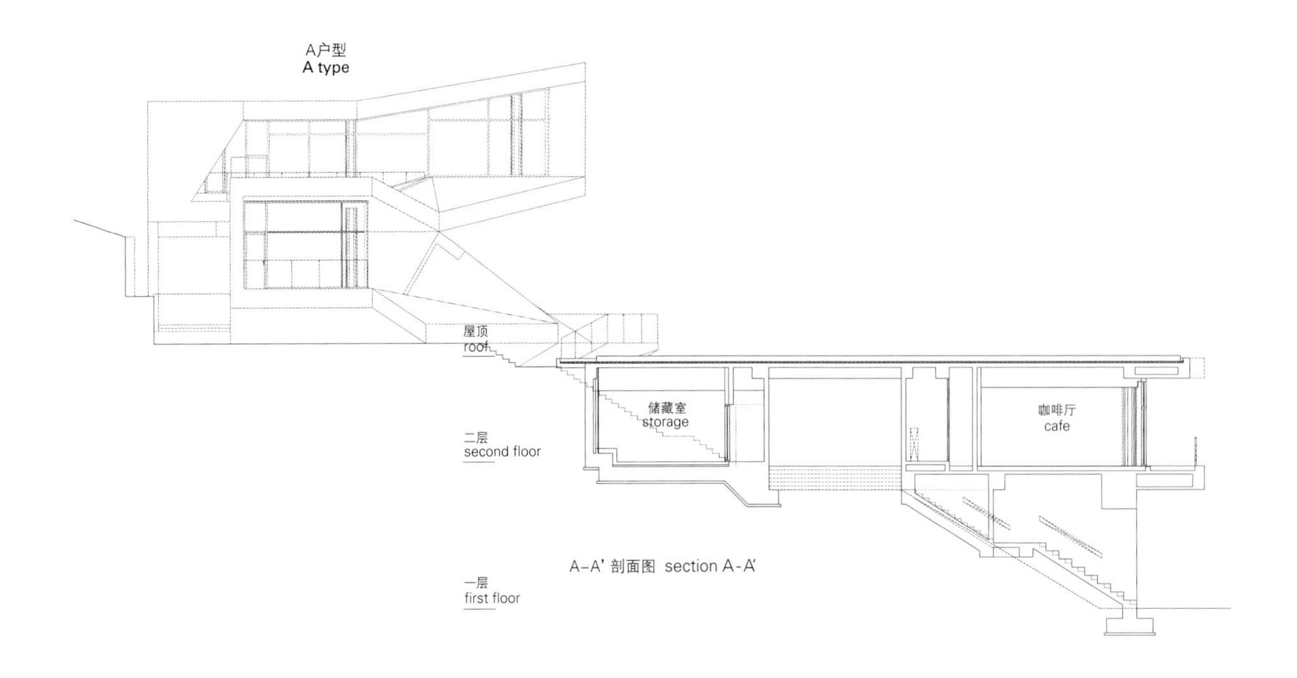

A–A' 剖面图 section A-A'

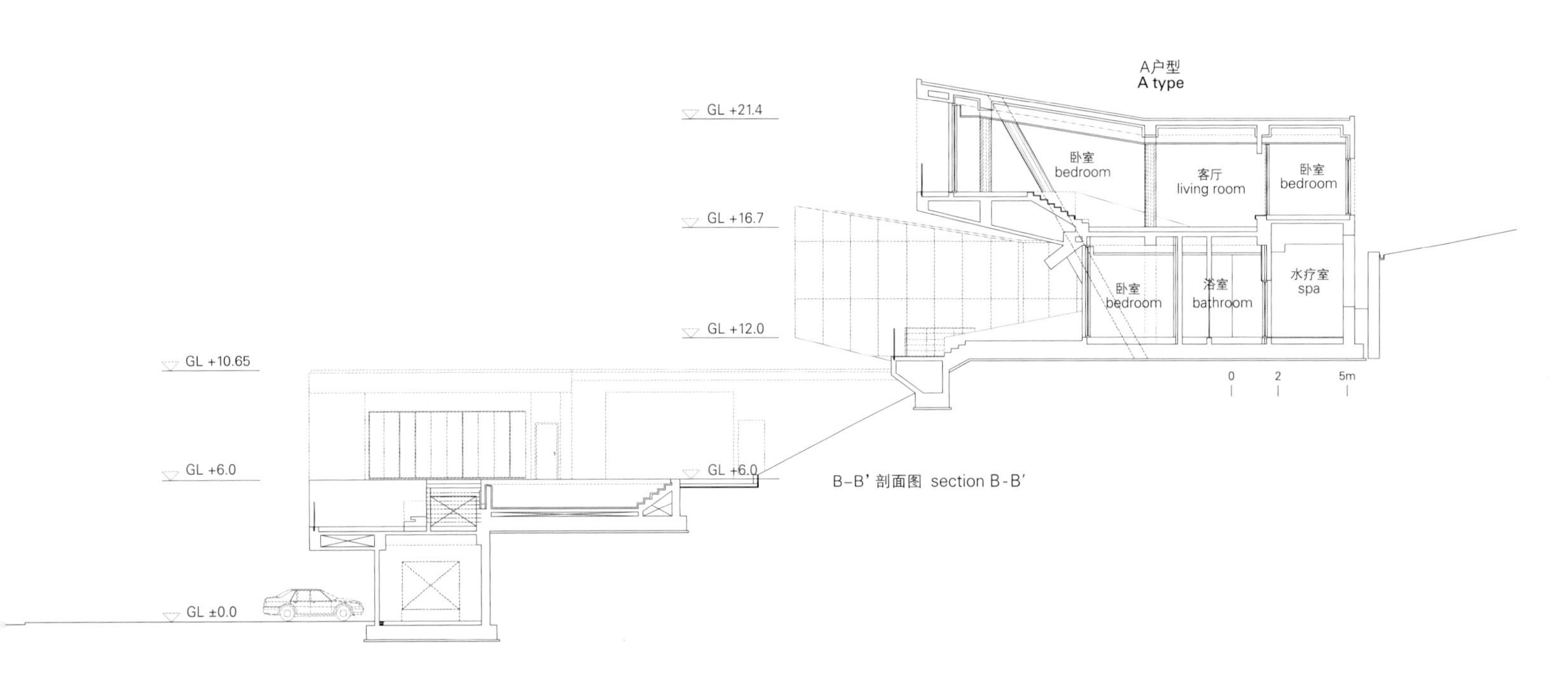

B–B' 剖面图 section B-B'

项目名称：U RETREAT
地点：HongCheon-gun, GangWon-do, Korea
建筑师：HeeSoo Kwak_IDMM Architects
用途：accommodation
用地面积：4,929.00m²
建筑面积：1,221.84m²
总建筑面积：1,595.29m²
建筑规模：two stories above ground
高度：22.2m
结构：RC
外部装饰：exposed concrete
竣工时间：2016.4
摄影师：©Kim Jaeyoun (courtesy of the architect)

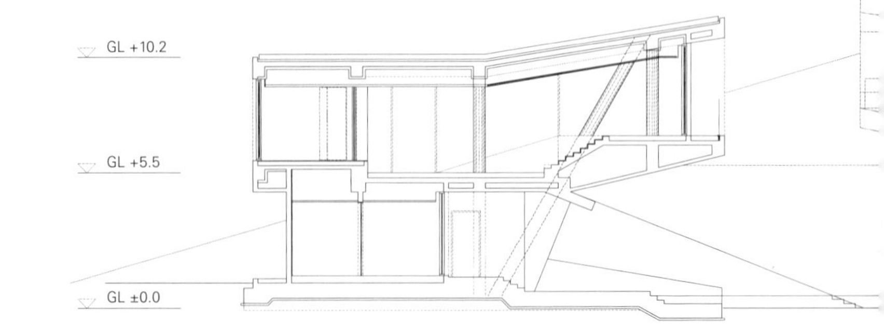

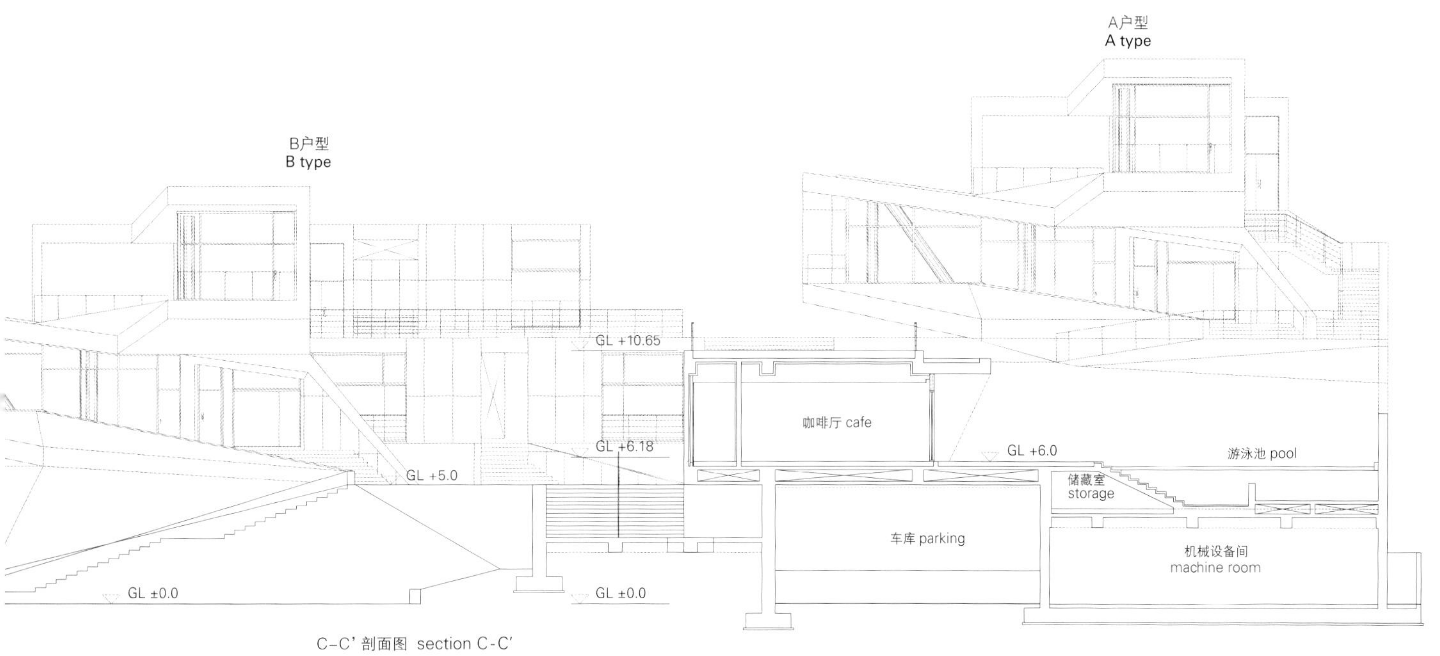

C-C' 剖面图 section C-C'

The wind passes through the valley, and the entire cliff sways with movement. The small gestures of the trees, each of which shakes according to its weight, come together forming a flowing cliff. How could the climate, as well as the movement and flow of the cliff changing with each and every instant, be interpreted and embodied in architecture?
The project was asked by a retired business man who had long been involved in fashion industry. After retirement, he wanted to host a unique meditative accommodation located deep in a valley. The important point of the project was to evolve the so-called "Pension", a type of small local accommodations, into a special place. The architect individually interpreted nature and analyzed diverse views from the position of the guests to provide a new spatial experience.
Guests vary from families for one night trip to couples for a date. All of them would want to take a rest at a fancy space without any disruption. They can experience a new spatiality where the living room, main bedroom, Korean-style bedroom, bathroom, terrace and spa are divided through different levels rather than are partitioned on one floor. The reason for this multi-leveling is that various levels can offer diverse relationships between the guest and the natural scenery, enabling a

vibrant appreciation for nature.
The mountain viewed from the site delivers the image of flying solid mass. Also oblique lines made by concrete mass are expected to be able to contrast with irregular and varied lines of nature. In this respect, diagonal columns and cantilevers, upper and lower parts connected with a twist generate an active shape change, in harmony with the natural scene. It deals with the hilly land through minimizing contact with the surface. This is an incidental result of interaction between the cantilevering masses and the undulating landscape.
U Retreat faces a cliff on the opposite side. As if the cliff were a background that alters with the season, U Retreat naturally embraces colorful scenes of the cliff via glass windows and multi-leveled platforms of concrete.
U Retreat has two types of units : upper(108m²) and lower(49m²) units. The lower unit faces the ground and is separated from the upper unit although they are linked within one mass. The upper unit has a sky terrace and spa where the guests can enjoy meditative bath viewing the cliff and sky.
Utility spaces include concierge, cafe, restaurant with a small kitchen. They support outdoor activities within U Retreat such as swimming, barbecue, and meditation.

FU 住宅

Kubota Architect Atelier

日本山口县的周南市是濑户内海周边一系列工业区的一个，而濑户内海海岸线充斥着众多的工业建筑。对于在这个地方长大的我而言，刺眼的金属油罐和凶猛的火势划过夜空的景象已是司空见惯了。这是一个超现实的、抽象的世界，让我们瞥见了未来的曙光，期待着一个新时代的到来。

项目场地距濑户内海不足一公里。新开发的城市住宅区密集地分布着新建房屋，很难欣赏到美景或无污染的大自然。而且，建筑场地邻近主干道，既嘈杂又喧闹。因此，本项目的目的是为客户及其家人建造一座能够开创一个崭新未来的建筑，他们将居住在这里，并在未来的多年生活中建立属于自己的生活方式。

在这里，光和影、风和水都不断地变得模糊抽象。三块白色厚板，每一块都被折成 L 形，以三维立体的形式嵌入这一不断变化的环境中。在交汇处所创造的空间是自然与人类情感之间的催化剂，同时也通过折叠的 L 形控制着自然的动态活力。阳光和风渗入或穿过厚板之间的空间。经过这些厚板的漫射，微光和涌动的水流产生了反射的效果。而透过这些现象，我们可以感知到一个广阔世界的存在，它超越了我们肉眼可见的微弱的自然痕迹。大自然包含了无限的理念，如果再融合人类情感的话，那么这两者就会共享无限的自由，为家庭提供一个无限开放而又丰富的未来生活。

FU-House

Shunan City, Yamaguchi is one of a series of industrial zones situated around the Seto Inland Sea, where the coastline bristles with industrial complexes. Having been raised in the area, I'm familiar with the sights of the glaring metal tanks and the fierce fire tearing through the dark night. It is a surreal and abstract world, which gives us a glimpse of the future glimmering with hope for the coming of a new era.
The project site is located less than a kilo-meter away from Seto Inland Sea. The recently developed residential district is densely distributed with new-built houses, and is far from being privileged to beautiful scenery or a healing nature. Fur-

项目名称：FU-HOUSE
地点：Yamaguchi, Japan
建筑师：Kubota Architect Atelier
主管建筑师：Katsufumi Kubota
项目团队：Kazuya Toizaki, Kazusa Kubota, Ichiki Ushirodani
顾问：Masayoshi Nakahara
结构工程师：Kenji Nawa _ Nawakenji-m, Nobuyuki Morinaga
总承包商：Jun Tokimori, Hiroaki Ueyama, Akinobu Tanaka, Yuka Housako _ Architects Studio Japan / TOKIMORI KENSETSU
用地面积：217.15m²
建筑面积：90.77 m²
总建筑面积：148.53m²
建筑覆盖率：41.8%
总楼面比率：56.57%
建筑规模：two stories above ground
结构：reinforced concrete
外部装饰：acrylic rubber type waterproof paint, compound cement, exposed concrete
内部装饰：exposed concrete
设计时间：2012—2014
施工时间：2015—2016
摄影师：©Kenji Masunaga (courtesy of the architect)

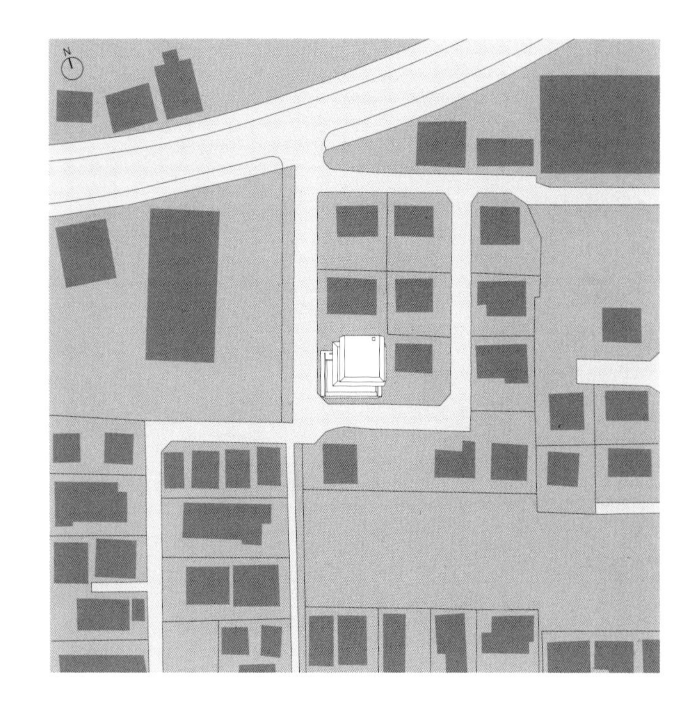

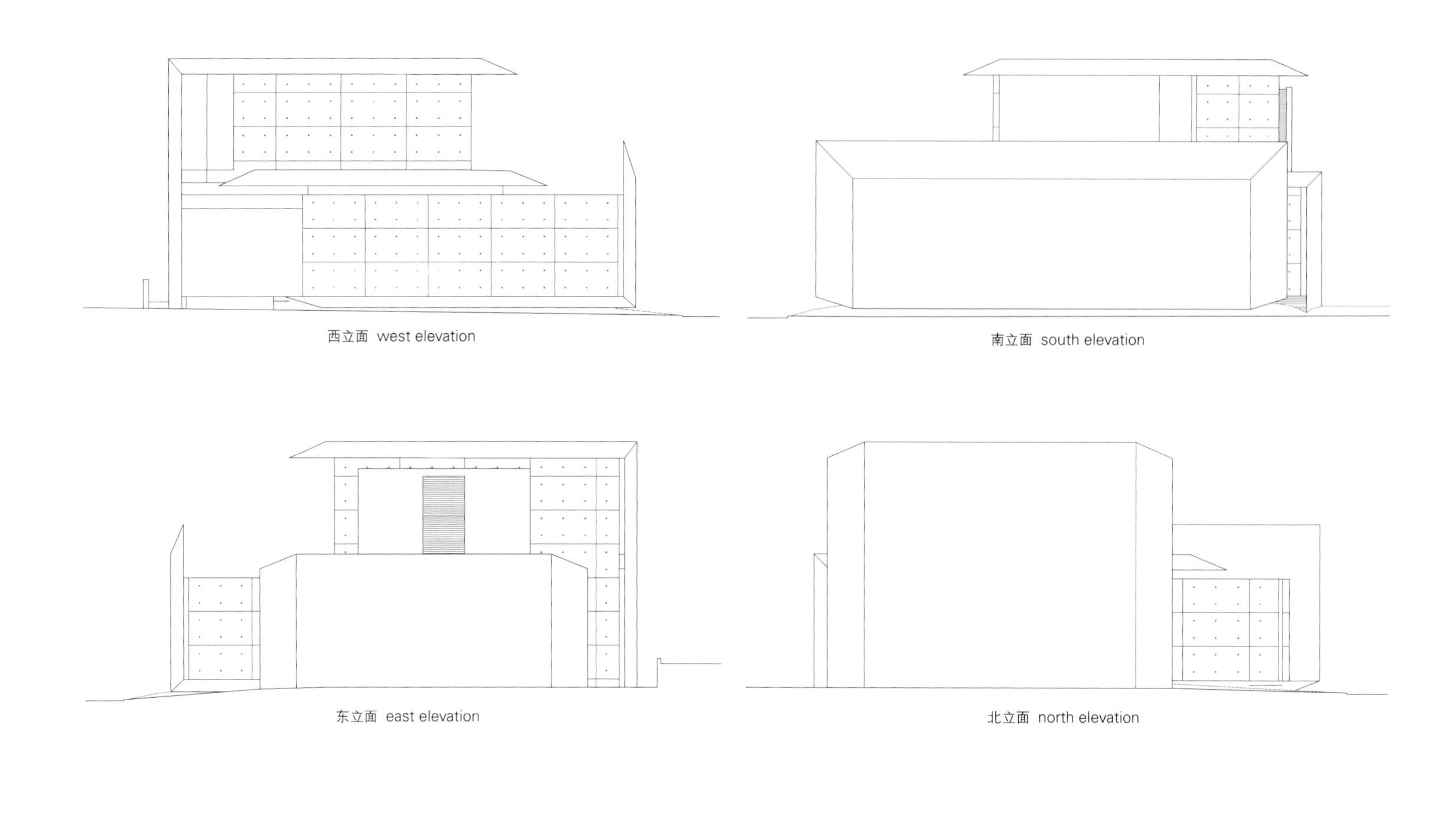

西立面 west elevation

南立面 south elevation

东立面 east elevation

北立面 north elevation

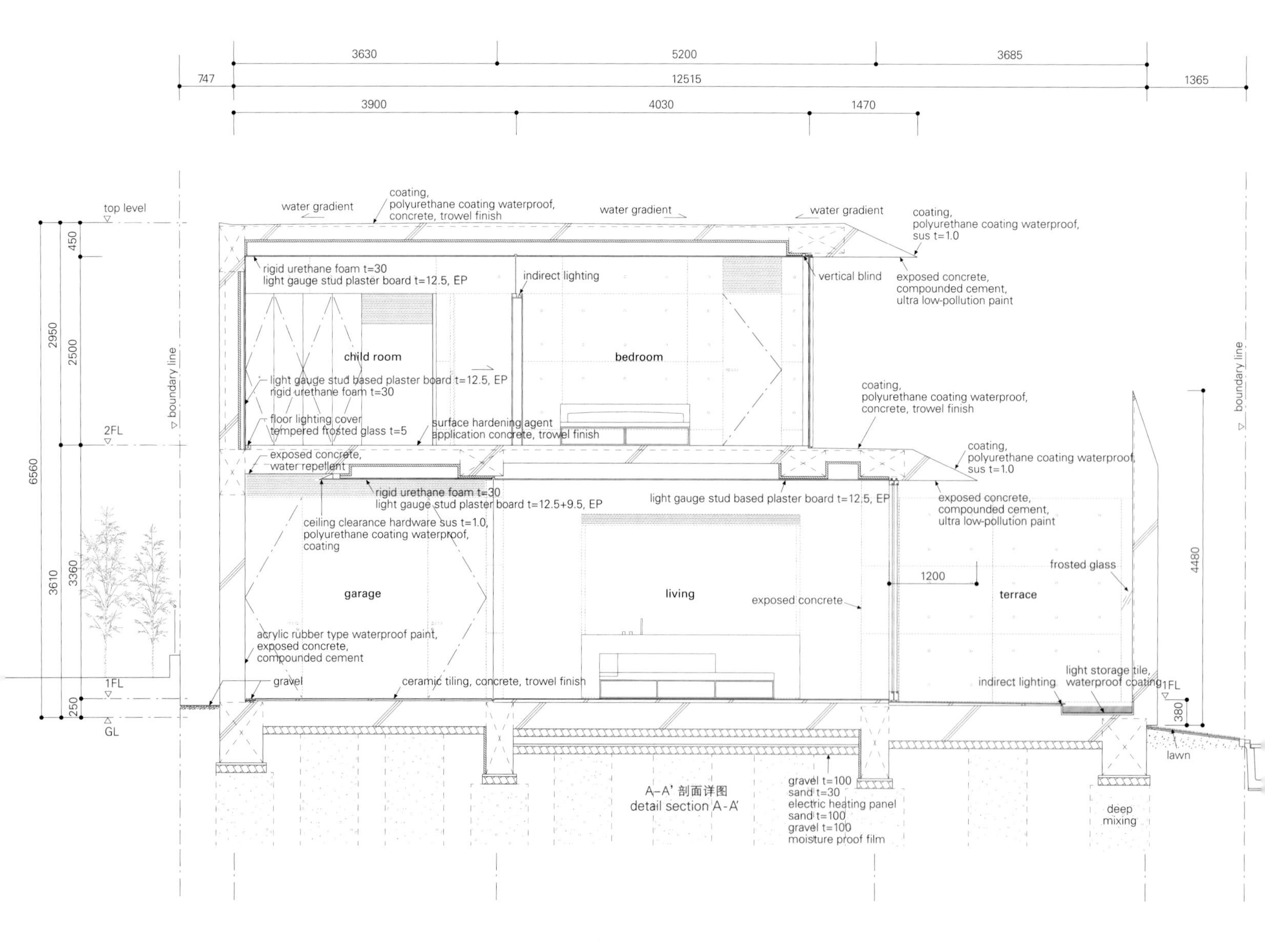

A–A' 剖面详图
detail section A-A'

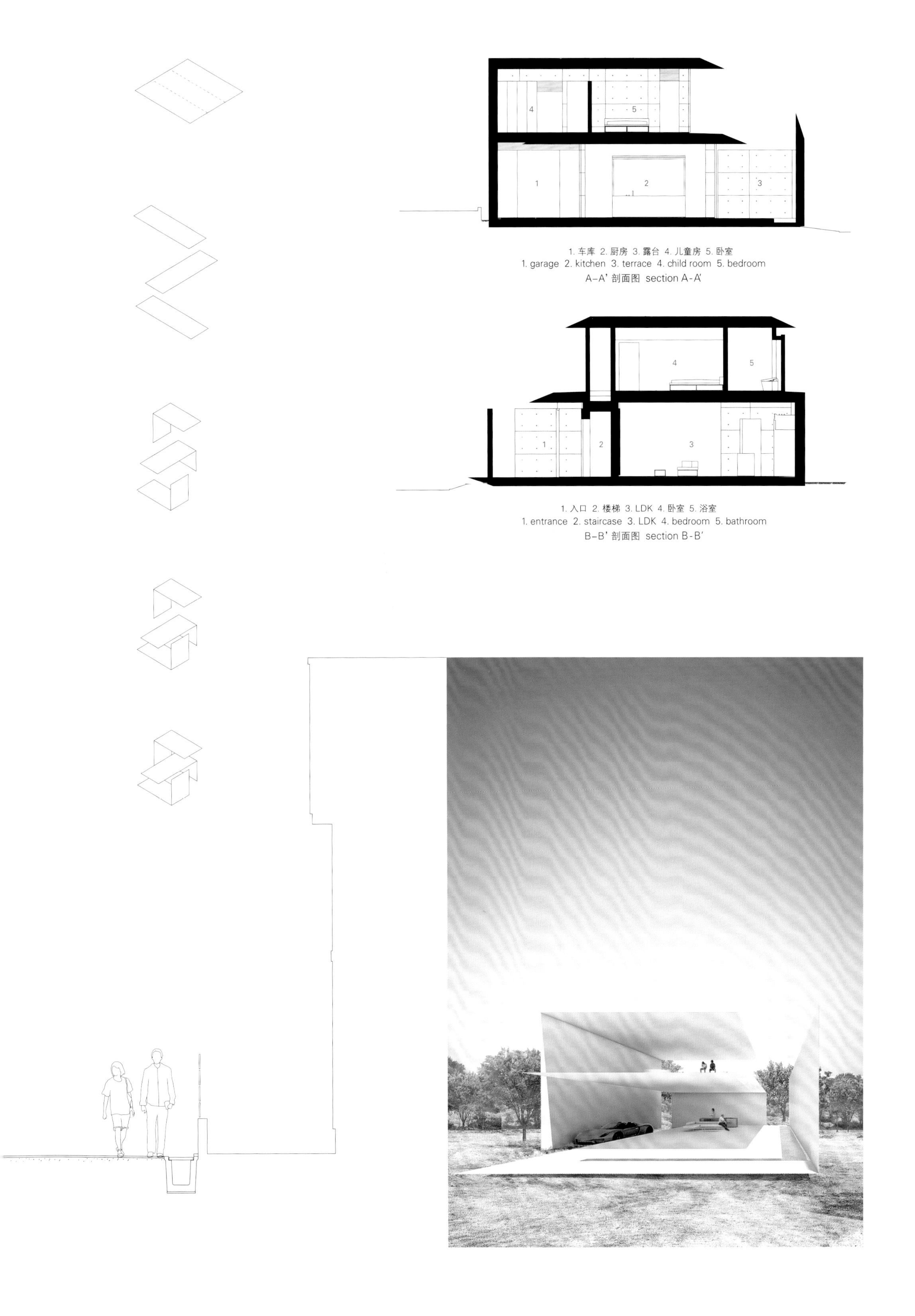

1. 车库 2. 厨房 3. 露台 4. 儿童房 5. 卧室
1. garage 2. kitchen 3. terrace 4. child room 5. bedroom
A–A' 剖面图 section A-A'

1. 入口 2. 楼梯 3. LDK 4. 卧室 5. 浴室
1. entrance 2. staircase 3. LDK 4. bedroom 5. bathroom
B–B' 剖面图 section B-B'

thermore, a major road passes nearby, causing tremendous noise and bustle. Here, the aim is to create an architecture that is able to open up a new future for the client and his family, who will settle here and establish their way of living in the many years to come.

Light and shadow are restlessly ambiguous and abstract; as are wind and water. Three pieces of white slab, each folded into an L-shape, are inserted three-dimensionally into this ever-changing environment. Spaces created between the intersections function as a catalyst between nature and human emotions, whilst also controlling nature's dynamism with their L-shape folds. Light seeps in and wind passes through the spaces between the slabs. Pale light and streaming water are diffused by the slabs acting as reflectors. These phenomena are indications of the existence of a vast world that extends beyond the faint traces of nature visible to our eyes. Nature embraces the idea of infinity, and by infusing it together with human emotions, the two share an unrestricted freedom, offering a boundlessly open and rich future for the family.

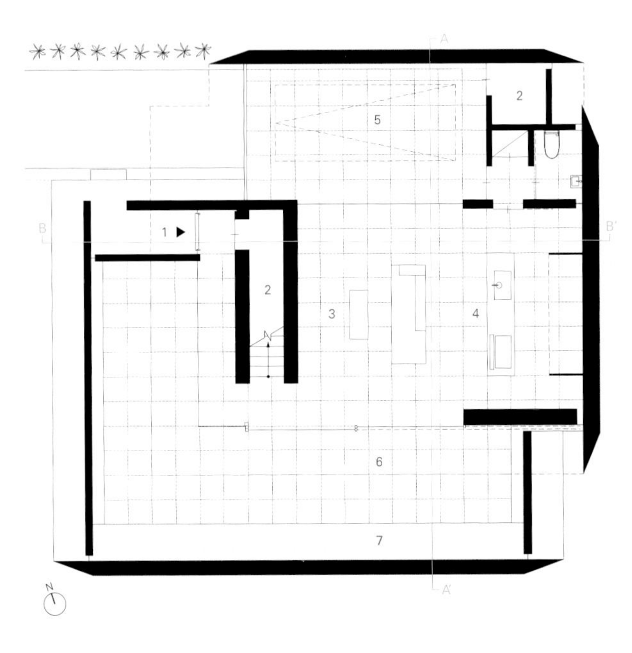

1. 入口 2. 楼梯 3. 客厅 4. 餐厅/厨房 5. 车库 6. 露台 7. 游泳池
1. entrance 2. staircase 3. living room
4. dinning/kitchen 5. garage 6. terrace 7. pool
一层 first floor

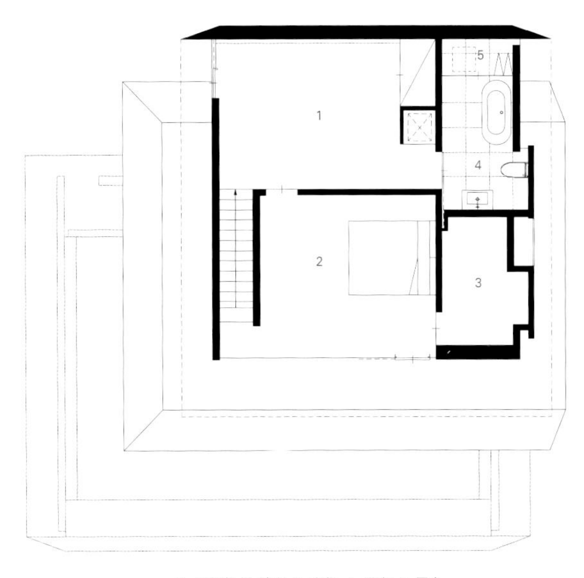

1. 儿童房 2. 卧室 3. 衣橱 4. 浴室 5. 露台
1. child room 2. bedroom 3. wardrobe 4. bathroom 5. terrace
二层 second floor

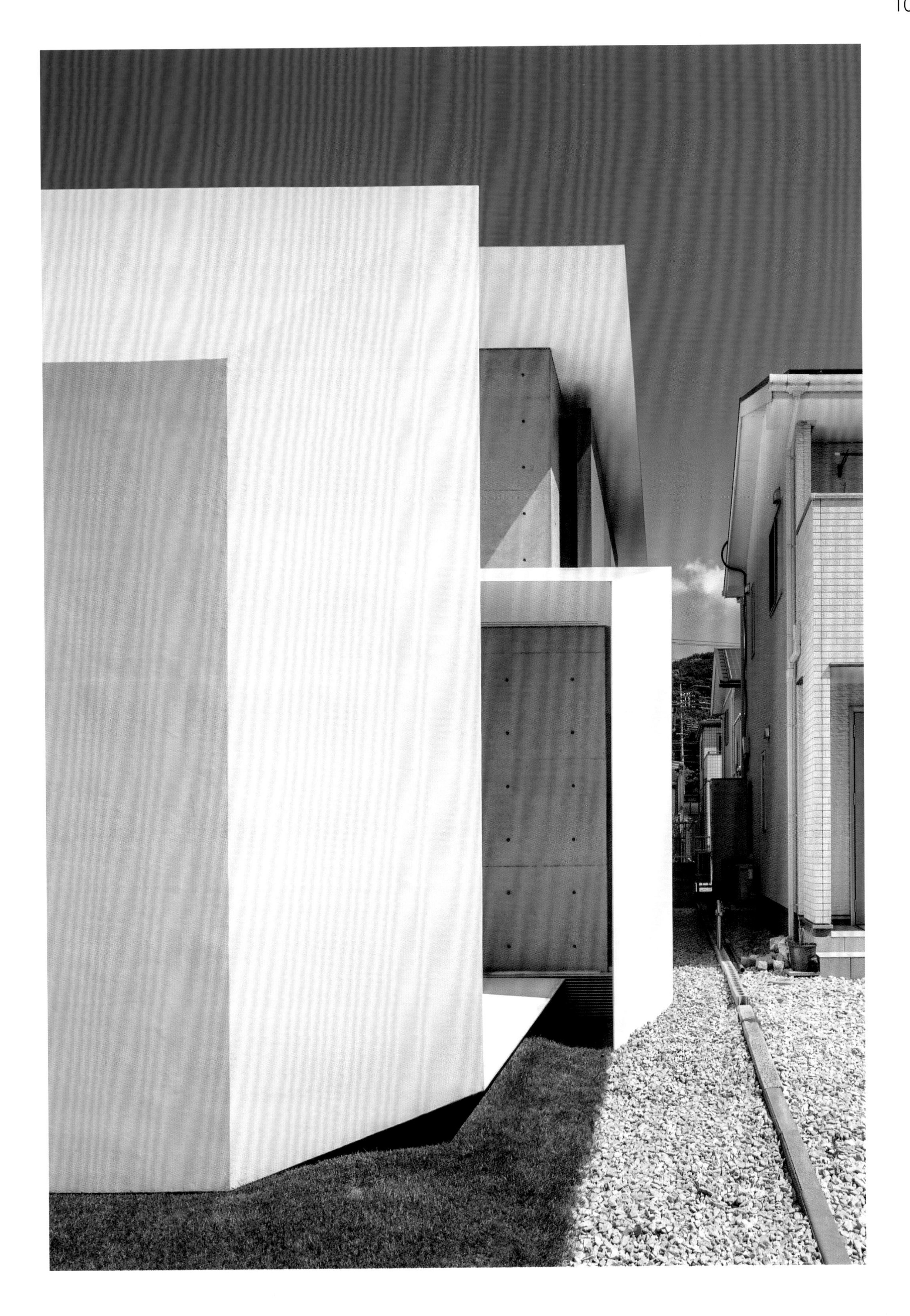

上海西岸 Fab-Union Space 建筑

Archi-Union Architects

在快速的城市化建设过程中，微型建筑的设计往往会带有许多不确定性。它既要求建筑师实现最大化的建筑容积率，充分利用空间，同时还要求创造一个可以适应未来多种使用可能的通用空间。同时，该项目的空间设计还必须从周边环境中脱颖而出。

Fab–Union Space在未来将成为一个非营利性的当代艺术、建筑和文化交流中心。其设计目的是成为一个展览和交流空间。项目位于上海的西外滩区域，其规划目的是成为上海的一个未来艺术和文化中心。尽管FU Space建筑的面积很小（大约300m²），但它是对数字化设计施工过程的又一次尝试。它体现了建筑师对于空间、材料和施工方法的新探索和新思考。

在设计之初，为了最大化地利用空间，建筑师将整个项目纵向划分为东西两部分，东侧为两层高度为4.2m的展厅空间，而西侧则为三层高度为2.8m的普通展厅空间。两侧的不同标高不仅提供了最大化的建筑容积率，也为办公和展览等未来的功能提供了空间的灵活性。外部两堵厚度为150mm的混凝土墙为楼板提供支承力。在空间的中部，楼板由楼梯支撑，而楼板的重力则是由这种有意而为之的交通流线设计实现的。在这里，传统意义上的交通空间成为结构，它突破了结构与交通流线之间的二元关系，而这两者又成为一种同化的建筑要素。

同时，交通流线和重力传导之间的相互制约和平衡形成了本案的空间形态设计。动态的非线性空间表面将重力分散传导，而不同的非线性表面又互相提供支撑。建筑采用了极简化的透明立面，并且在建筑外部就可以解读内部的混凝土结构。这样的设计策略确保了每一侧的展室都具备空间完整性，而交通空间的连结则强化了人在建筑中的动态行为。

而根据空气动力学原理，它使整栋建筑的通风达到了最大化。所有特别的空间体验都产生于中间的交通流线空间。遵从外部找形工艺的内部公共空间，在营造一种仿佛攀登中国古典园林中的假山的抽象思维过程中得到了增强。中国园林设计的关键因素在于以小见大，它在步移景异的情况下尤为有效。这种抽象的休验过程是通过从不同角度都得到了强化的高性能表面几何体实现的。

作为一种可塑材料，混凝土具备结构特性，并且具有外部施工的特点。动态非线性空间形体的建造建立在结构性能优化和空间动态学的基础之上。这整个过程运用了石料切割、透视几何和算法构型等众多设计方法。整座建筑从设计到施工历时仅四个月，它应该算是一种奇迹般的数字化设计及施工方法了。

Fab-Union Space

There are many uncertainties to design micro-scale architecture during the rapid Chinese urbanization process. It demands architects to maximize the floor area ratio, the full utilization of the space, and at the same time to create a universal space which can adapt to the future multiply programs. Meanwhile, the project has to be spatially identified from its surrounding context.

Fab-Union Space will become a future non-profit contemporary art, architecture and culture communication center. It is aimed to be an exhibition and communication space. It is located in the West Bund area in Shanghai, which is planned to be a future art and culture center of the city. Although it is micro in scale (around 300m²), FU Space is another parametric design construction practice. It represents the architects' recently research and currently new thinking about space, material, and construction methodology.

In the beginning of the design, in order to maximize the utilization of space, the project has been divided into two parts: west and east, the east part has two floors of 4.2-meter-

渐进结构优化法（BESO）
evolutionary structural optimization (BESO)

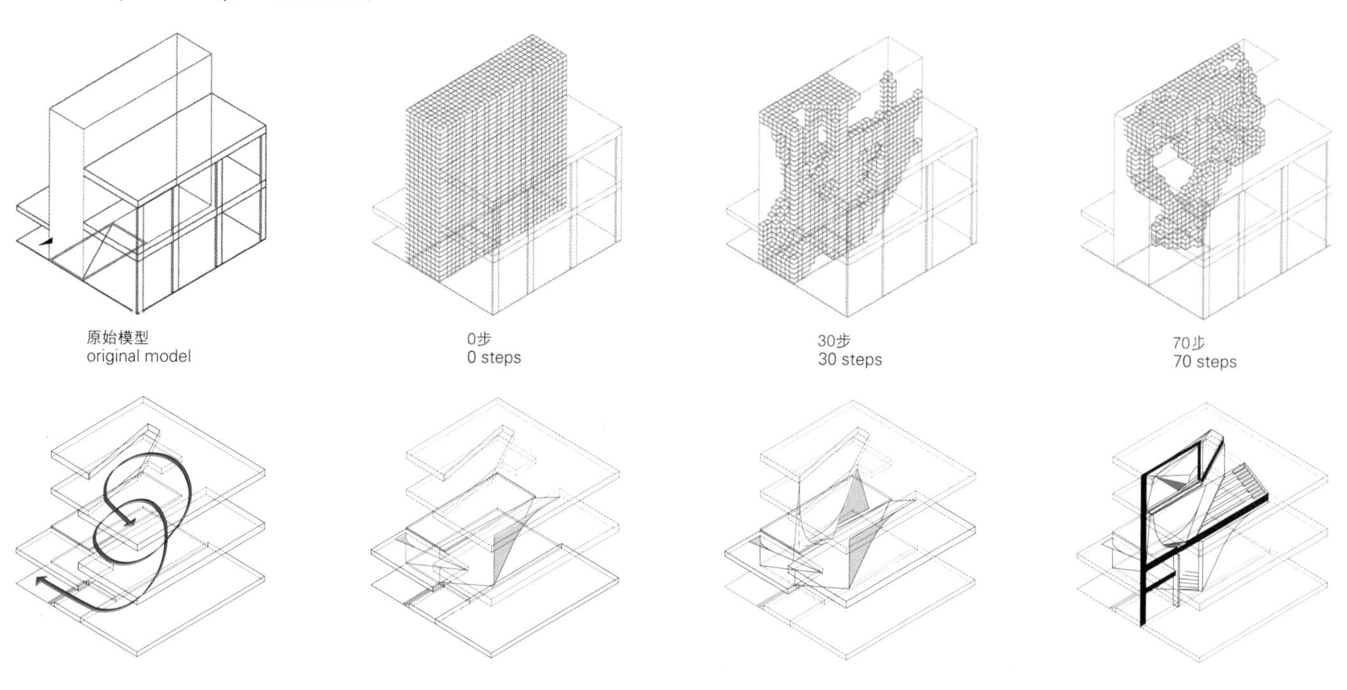

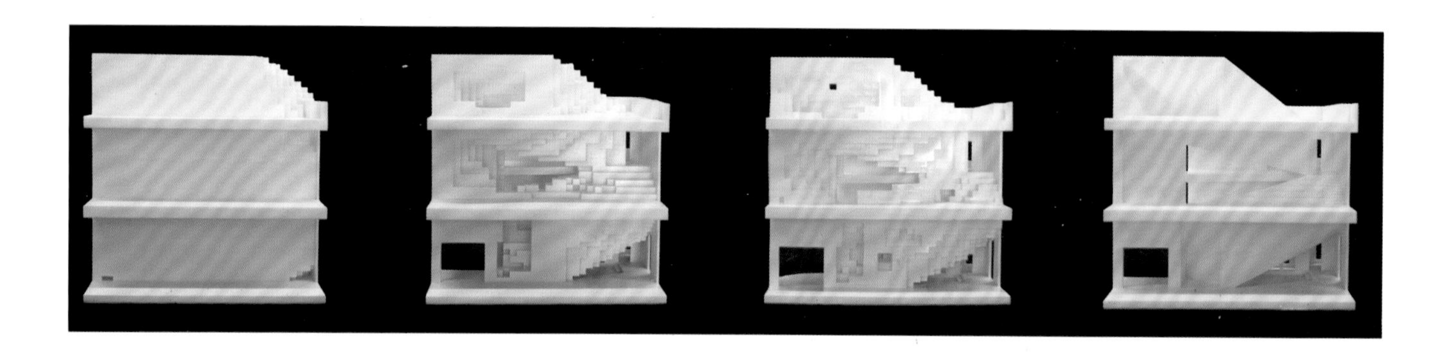

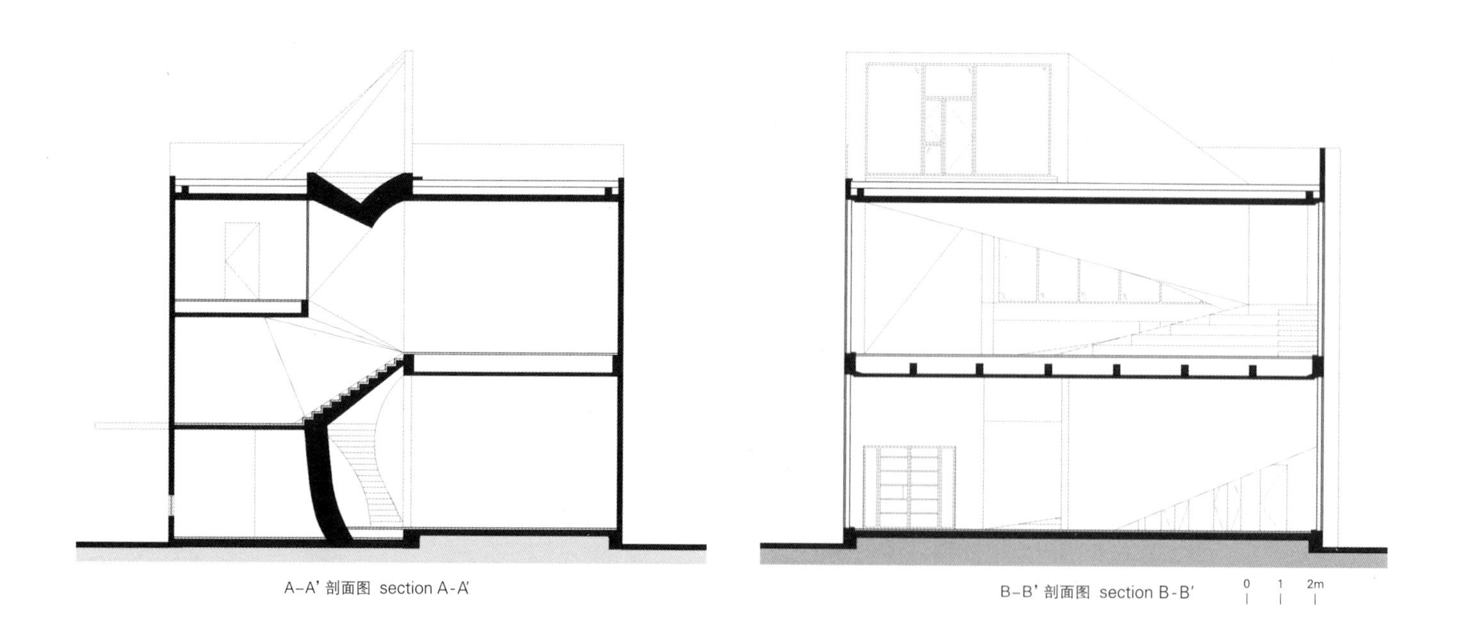

A-A' 剖面图 section A-A'

B-B' 剖面图 section B-B'

©Shangliang Su (courtesy of the architect)

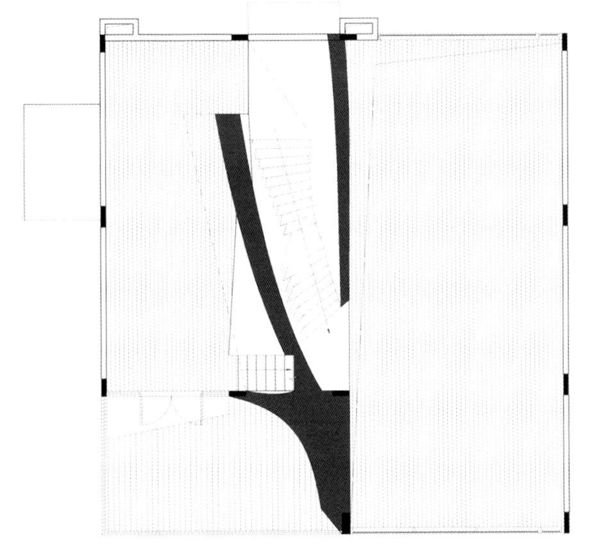

夹层 mezzanine (+3.4m)

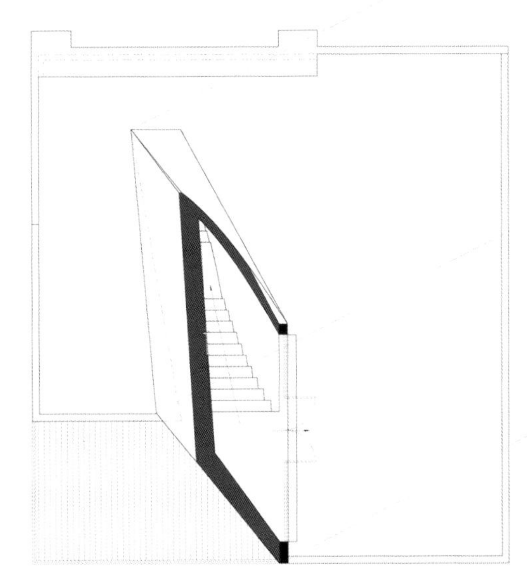

屋顶 roof

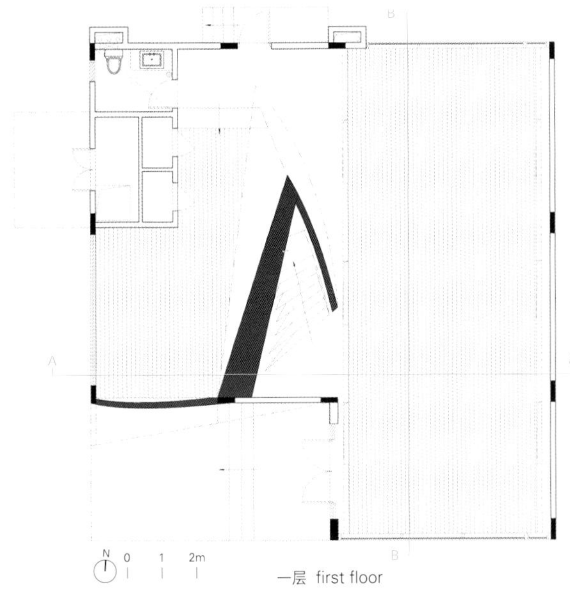

一层 first floor

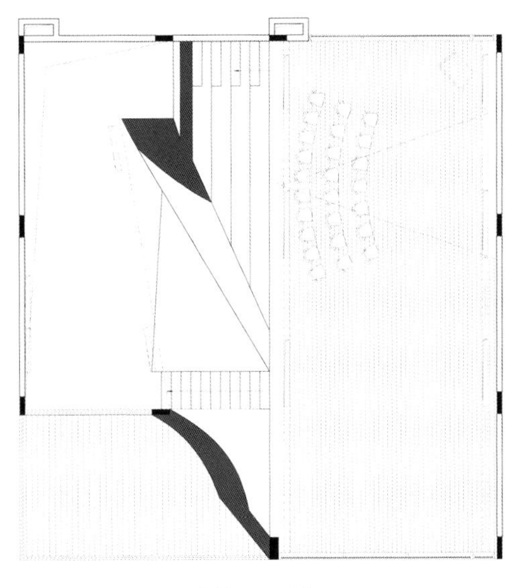

二层 second floor

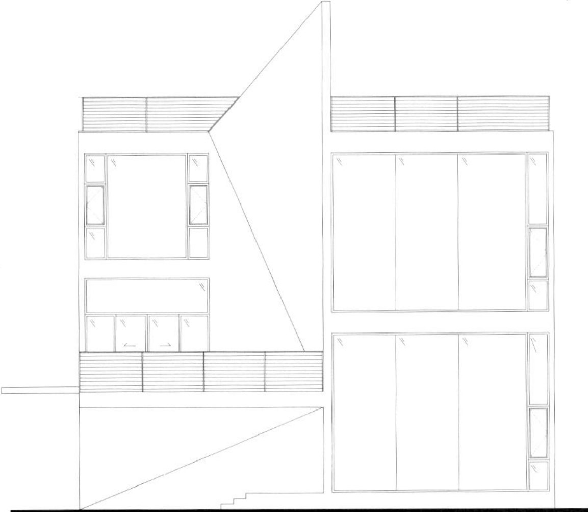

南立面 south elevation

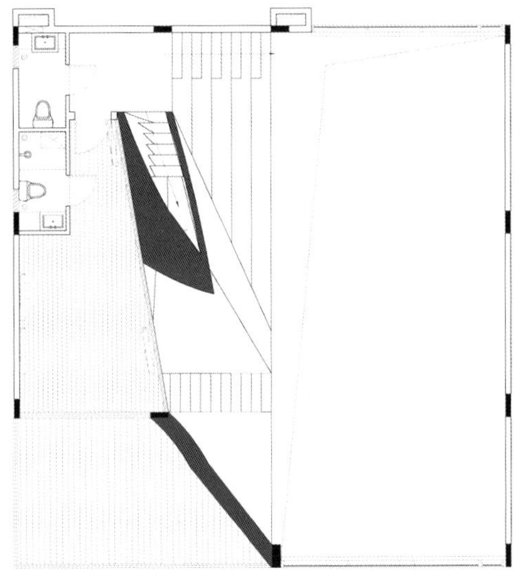

夹层 mezzanine (+6.93m)

high exhibition spaces and the west part has three floors of 2.8-meter-high regular exhibition spaces. The different heights between two sides not only provide the maximized floor area ratio, but also provide the spatial flexibility for the future office and exhibition programs. The floors are supported by the two 150mm concrete walls from the outside. In the middle of the space, and the floors are supported by the circulation stairs, the floor gravitational force has been carried out by the deliberate circulation design. At this point, the circulation becomes structure, which breaks the dualistic relationship between structure and circulation and they become one assimilated architectural element.

Meanwhile, the mutual restraint and balance between the circulation and the gravity conduction lead to the spatial form design. The dynamic nonlinear spatial surface conducts the gravitational force and different nonlinear surfaces support each other. The architectural façade is minimalistic pure transparency which externalizes the internal concrete structure from the outside. This design strategy ensures the spatial integrity of galleries on each side, and the connection circulation emphasizes the dynamic human movement.

And according to the aerodynamics theory it maximized the ventilation of the whole building. All special space experience lies in the in-between circulation space. Interior public space followed by the exterior form-finding process, is enhanced by an abstract thinking on creating a kind of experience climbing the rockery of Chinese garden. The key aspect of Chinese garden design is to make it big through a small scale, which is extremely efficient in changing sceneries with varying viewpoints. The abstract process for this kind of experience is achieved by the HP surface geometry, which is intensified from different perspectives.

As a plastic material, the concrete has the characteristics of construction and has the characteristics of external construction. The dynamic nonlinear spatial shape is built on the basis of structural performance optimization and spatial dynamics. The whole process uses a variety of design methods of cutting stone, perspective geometry, and the Algorithm configuration. The whole building from design to construction lasted for only four months should be a digital design and construction method of the miracle.

项目名称：Fab-Union Space on the West Bund / 地点：Building D, Longtengdadao, Xuhui District, Shanghai / 建筑师：Philip F. Yuan _ Archi-Union Architects
设计团队：Alex Han, Xiangping Kong, Xuwei Wang / 结构工程师：Zhun Zhang / 照明设计顾问：Guojian Hu, Linhua Yang / 立面设计顾问：Shanghai dimon curtain wall engineering co., LTD / 用地面积：245m² / 建筑面积：368m² / 总建筑面积：312m² / 结构：concrete shear wall / 施工时间：2015.6—9 / 竣工时间：2015
摄影师：©Hao Chen (courtesy of the architect) (except as noted)

Bosjes 教堂

Steyn Studio

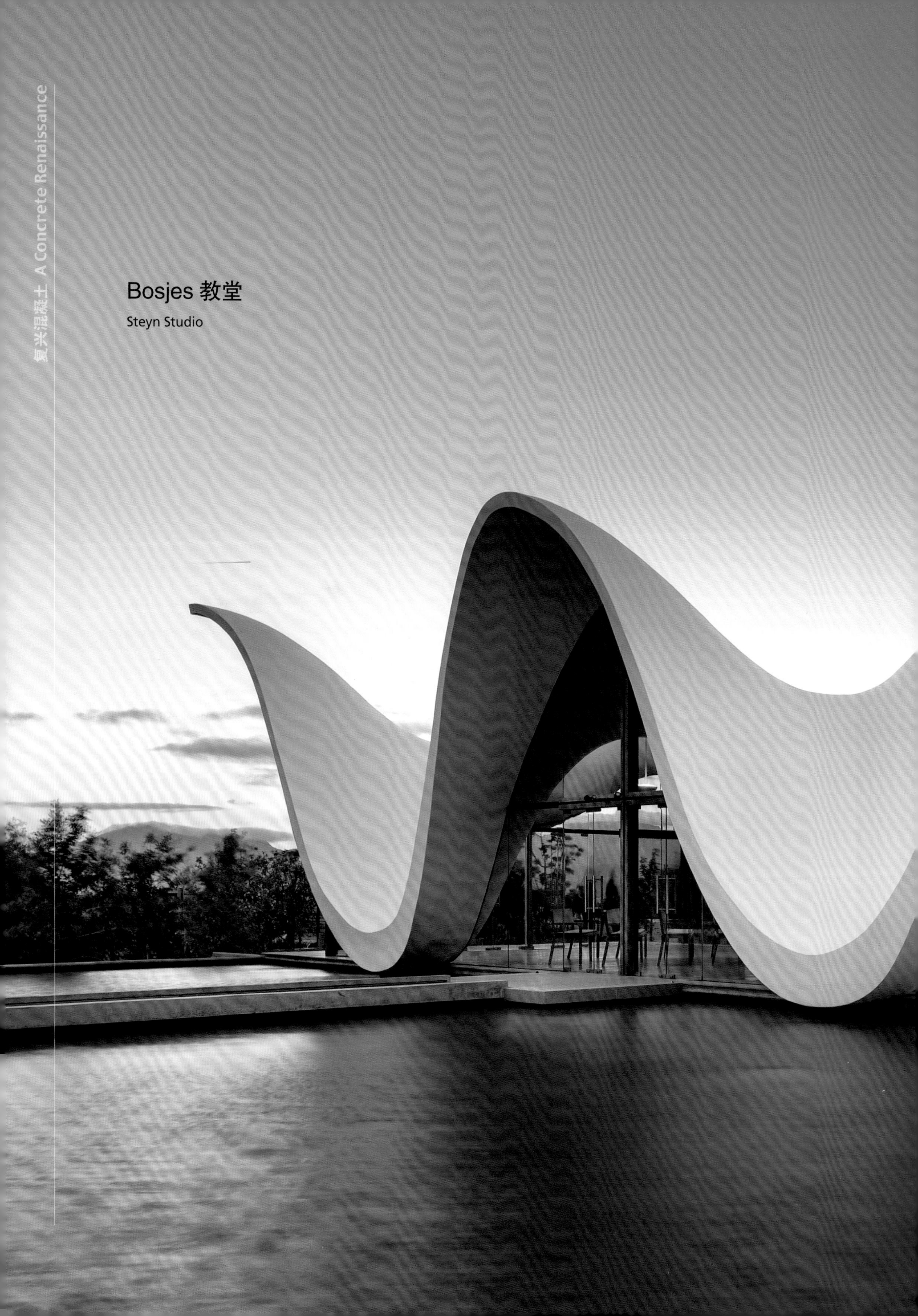

这座新教堂位于南非的一处葡萄园内，由伦敦斯泰恩工作室的库切·斯泰恩（出生在南非）担任设计。教堂平静的雕塑式外形模仿了附近山脉的轮廓，向点缀在西开普乡村景观中的历史上著名的开普荷兰风格山形墙致敬。屋顶由一个超薄的混凝土浇筑外壳构成，独自承重，每个波形轮廓都引人注目地连着地面。在屋顶结构的每个波浪的最高点位置，中央安装了十字架的大面积玻璃幕墙都起到了装饰立面的作用。

受到《诗篇36：7》的诗意灵感启发，清晰的白色形式被构思为一种轻盈而富有活力的结构，仿佛漂浮于山谷之中。波光粼粼的水池更使得建筑结构给人一种失重的错觉。教堂从它所在的平地上拔地而起，建立在一个基座上，在其周围环境中提供了一个层次分明的焦点。包括葡萄园和石榴果园在内的新绿化种植，将原本贫瘠的土地变成了一片郁郁葱葱的绿洲。

教堂内部采用简单的矩形平面，形成了一个大型的开放式组合空间。光滑的水磨石地板在室内反射光线。起伏的白色天花板投下了一连串的阴影，随着每日光线的变化，这些阴影像是在建筑内部"跳舞"。材料温和的色调为令人印象深刻的葡萄园及远山的景观创建了一个中性的背景。

为了保持简单的屋顶及组合空间的结构形式，建筑功能空间的其他构件不是隐藏于底座内，就是分散布置在周围花园的转角处。

受到19世纪在开普荷兰风格农场建造的摩拉维亚传教站的朴素外形的启发，教堂省去了尖顶——摒弃了与给人留下深刻印象的周围自然环境有关的重要意义。

这座教堂既具有开放性，也是一个向外延伸到山谷和山脉的空间，在周围环境中让人愈发意识到造物主的创造力。

Bosjes Chapel

The new chapel, set within a vineyard in South Africa, is designed by South-African born Coetzee Steyn of London-based Steyn Studio. Its serene sculptural form emulates the silhouette of surrounding mountain ranges, paying tribute to the historic Cape Dutch gables dotting the rural landscapes of the Western Cape. Constructed from a slim concrete cast shell, the roof supports itself as each undulation dramatically

项目名称：Bosjes Chapel
地点：Bosjes Farm, Witzenberg District, Western Cape, South Africa
建筑师：Steyn Studio, UK
项目建筑师：TV3 Architects, South Africa
家具设计：Liam Mooney Studio
承包商：Longworth & Faul
结构工程师：Henry Fagan & Partners
机械&电气工程师：Solution Station
工料测量师：De Leeuw
规划顾问：Ron Brunings
景观设计师：CNdV Landscape Architects
遗产顾问：Graham Jacobs, Elzet Albertyn & Lize Malan
总建筑面积：430m²
施工时间：2013—2016.12
摄影师：©Adam Letch (courtesy of the architect)

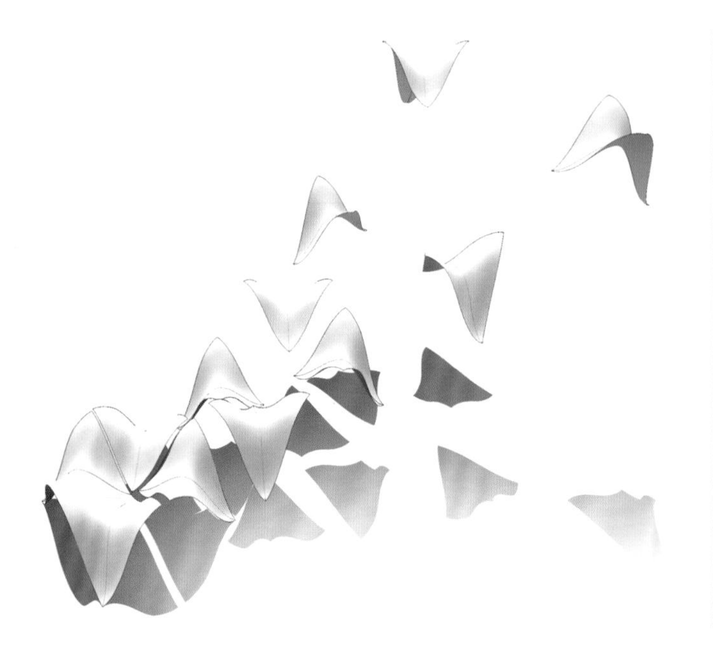

falls to meet the ground. Where each wave of the roof structure rises to a peak, expanses of glazing adjoined centrally by a crucifix adorn the facade.

Drawing poetic inspiration from Psalm 36:7, the crisp white form is conceived as a lightweight, and dynamic structure which appears to float within the valley. A reflective pond emphasises the apparent weightlessness of the structure. Elevated upon a plinth, the chapel rises from the flat land that it sits upon, providing a hierarchical focal point within its surroundings. New planting including a vineyard and pomegranate orchard creates a lush green oasis on the otherwise exposed site.

Inside, a large and open assembly space is created within a simple rectangular plan. Highly polished terazzo floors reflect light internally. The undulating whitewashed ceiling casts an array of shadows which dance within the volume as light levels change throughout the day. This modest palette of materials creates a neutral background to the impressive framed views of the vineyard and mountains beyond.

In order to keep the structural form of the roof and assembly space pure, other elements of the building's functional programme are either hidden within the plinth, or discretely within the outer corners of the surrounding garden.

Inspired by the simplicity of the Moravian Mission Stations established on Cape Dutch farms in the 19th Century, the chapel lacks a spire – relinquishing a sense of significance in relation to its impressive natural surroundings.

As an open embrace which invites in, the chapel is also a space that extends outwards into the valley and mountains beyond, raising the awareness of God's creation in the immediate environment.

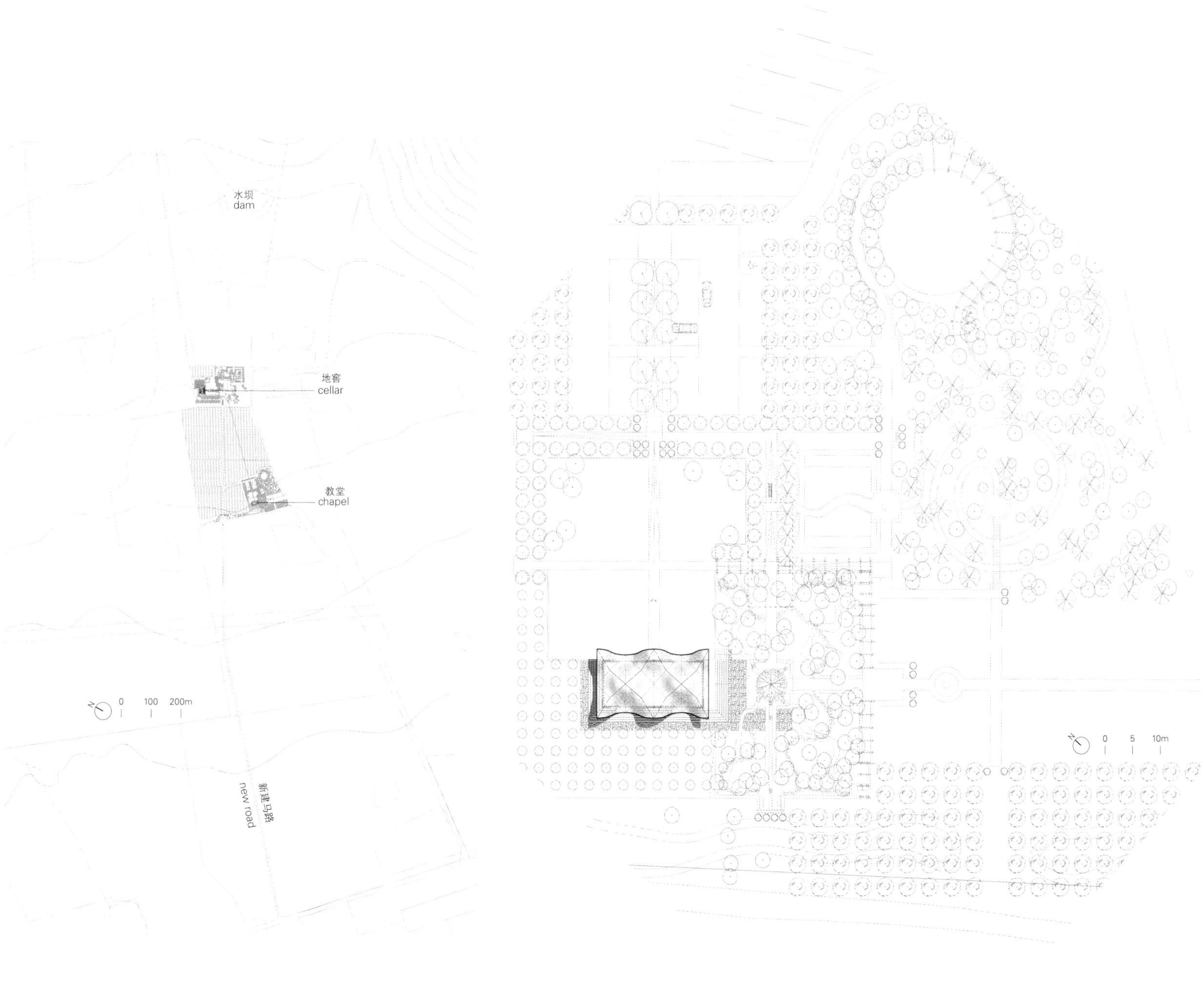

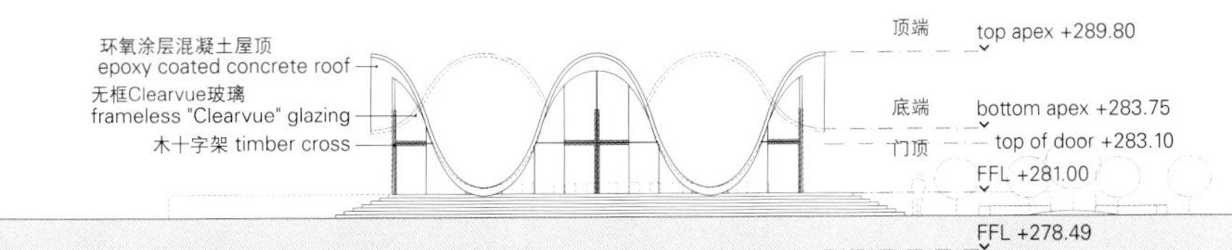

西南立面 south-west elevation

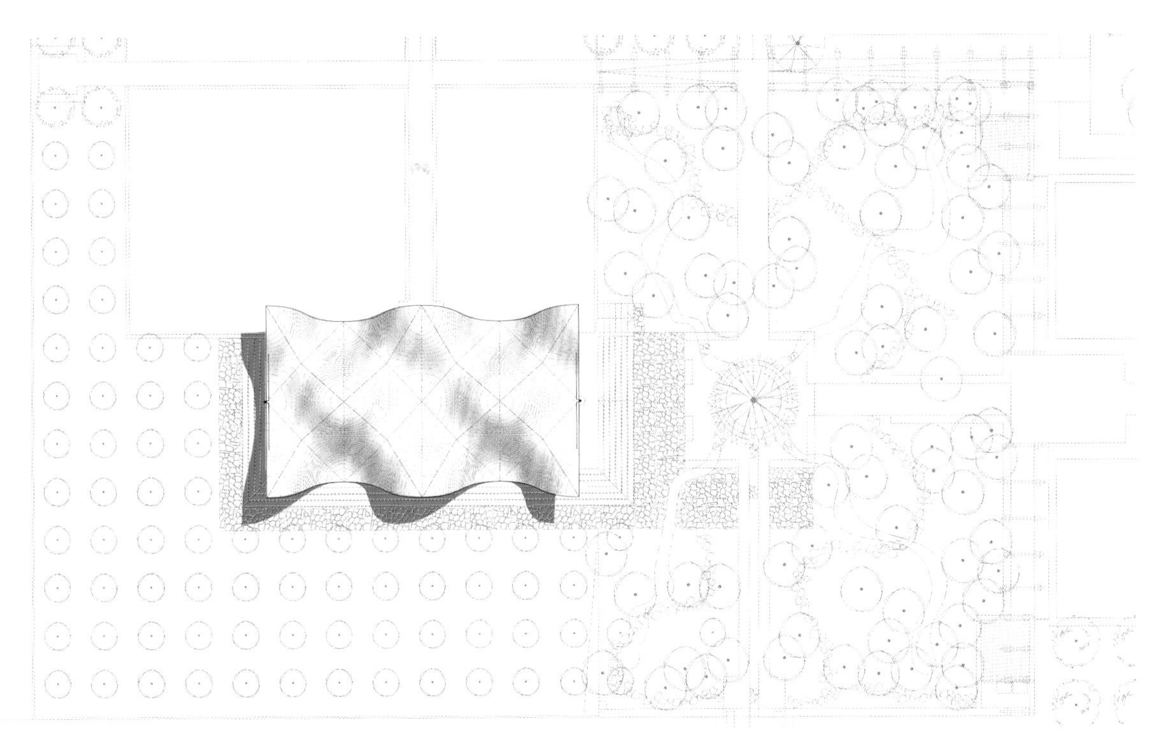

屋顶 roof

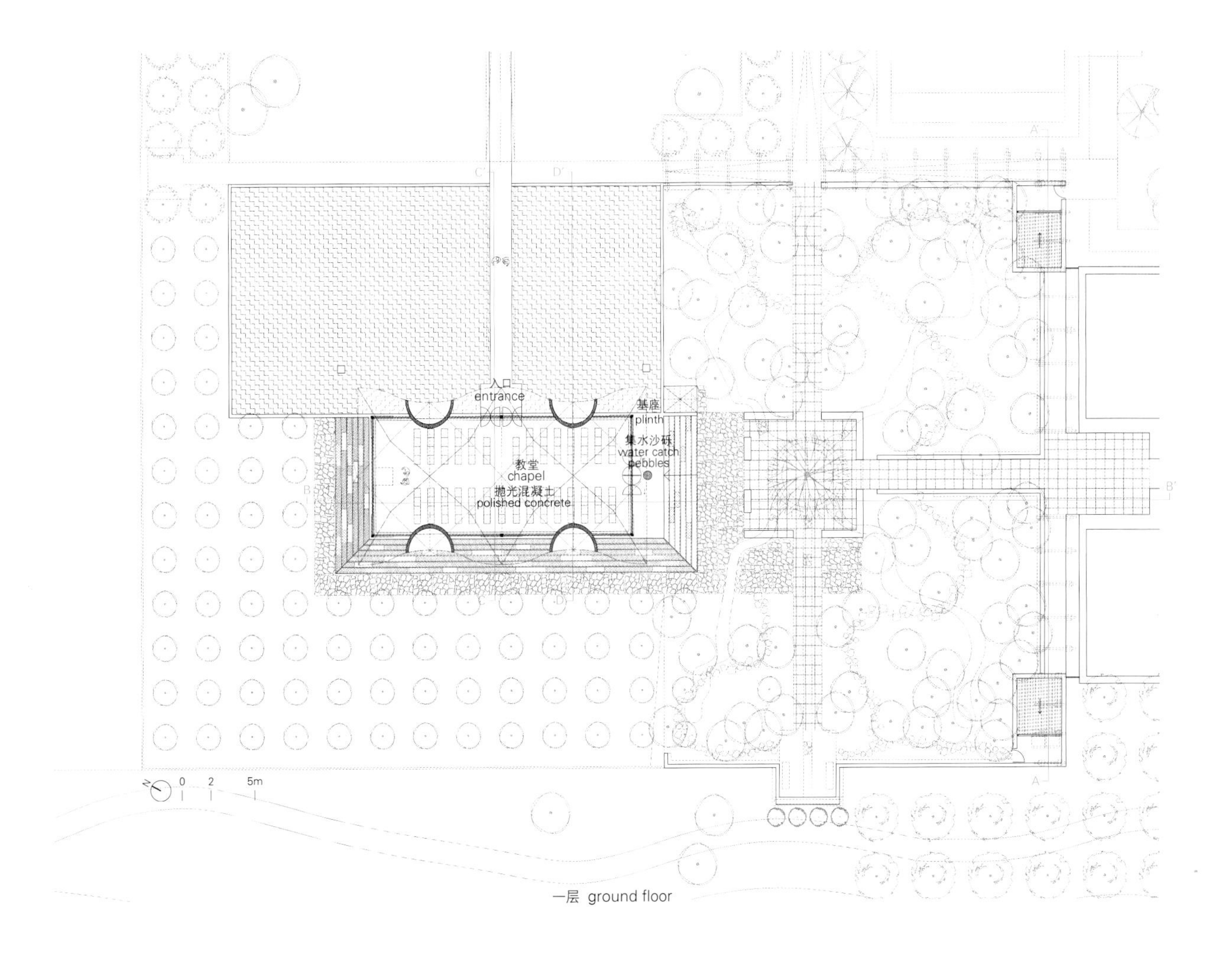

一层 ground floor

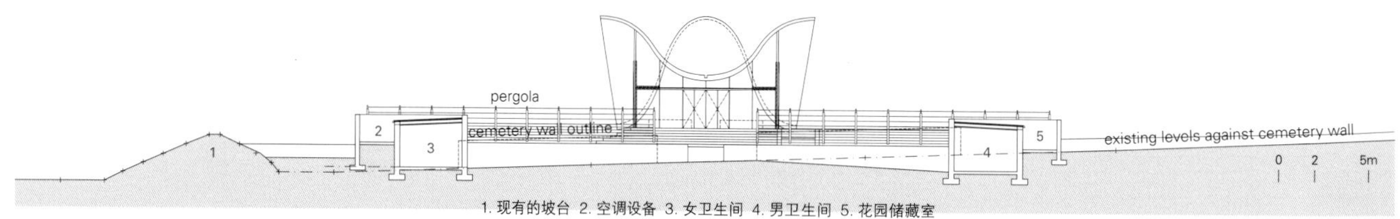

1. 现有的坡台 2. 空调设备 3. 女卫生间 4. 男卫生间 5. 花园储藏室
1. existing berm 2. A/C plant 3. female toilet 4. male toilet 5. garden storage
A–A' 剖面图 section A-A'

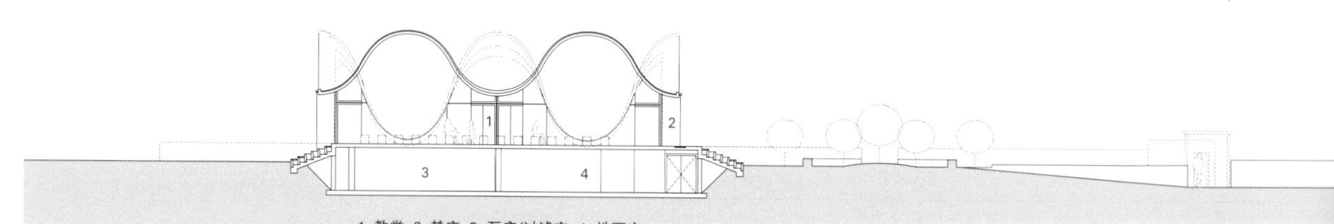

1. 教堂 2. 基座 3. 泵房/过滤室 4. 地下室
1. chapel 2. plinth 3. pump/filter room 4. basement
B–B' 剖面图 section B-B'

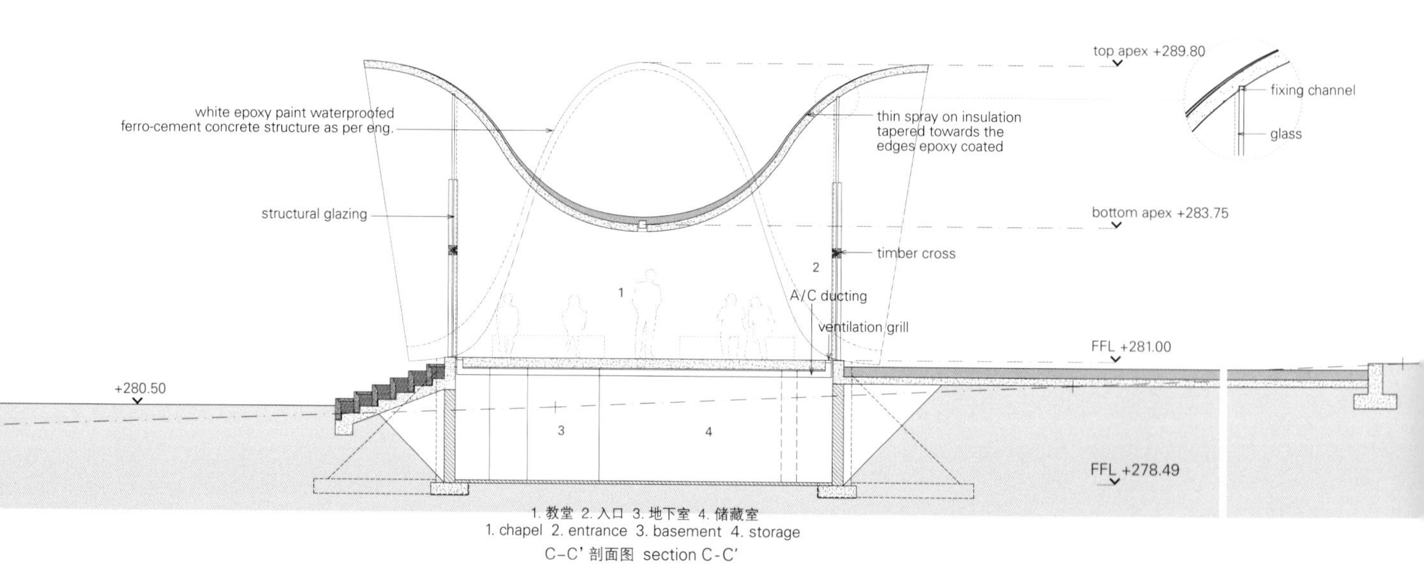

1. 教堂 2. 入口 3. 地下室 4. 储藏室
1. chapel 2. entrance 3. basement 4. storage
C–C' 剖面图 section C-C'

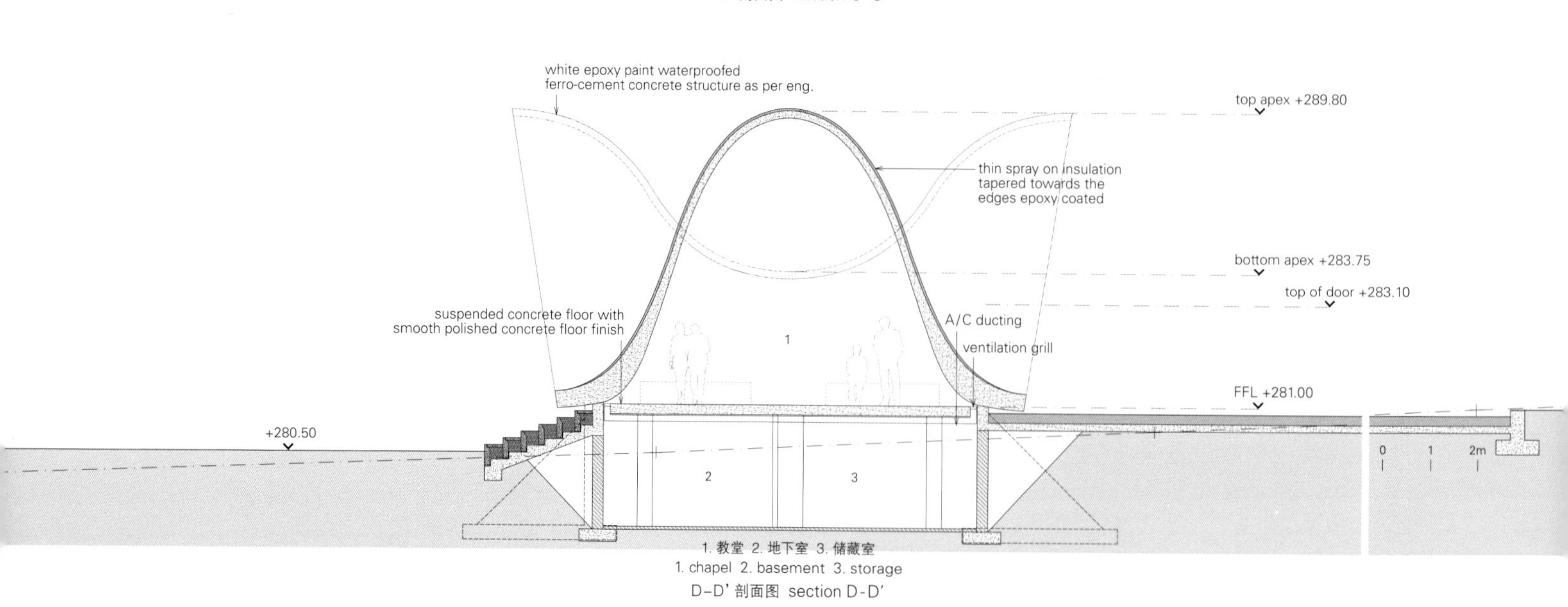

1. 教堂 2. 地下室 3. 储藏室
1. chapel 2. basement 3. storage
D–D' 剖面图 section D-D'

森林独奏屋

Office KGDVS

位于巴塞罗那南部马塔兰亚山区的森林独奏屋项目由法国开发商Christian Bourdais发起。这一批当代风格小型度假村原型由一群年轻的国际建筑师设计。

首栋建筑竣工于三年前，由智利建筑师二人组Mauricio Pezo和Sofia Von Ellrichshausen担任设计，而第二栋由比利时的Office KGDVS建筑师事务所设计，项目于4月24日启动。

第二栋森林独奏屋建造在高地的顶部，俯视着周围原始森林的风貌和大自然的鬼斧神工。因为这里的风景太让人印象深刻了，所以Office KGDVS建筑师事务所认为建筑应该在此隐匿，这样做只是希望更好地突出周围自然风光的特质。一个直径为45m的简洁圆形屋顶突出了高地及其边缘位置的特性。屋顶起到了建筑遮盖物的作用，并构成了居住区的边界。它由四排柱子提供支撑，这些排得笔直的柱子的位置刚好是圆形基座的弦切线。只有这四块带有多层防护的区域为居住空间。大型的自由伸缩幕墙可以在环形外墙面上滑行，使居住区能够完全敞开，从而使住户可以最大限度地接触周围的自然环境。因为该地块没有与任何设施相连，所以这座住宅是完全独立的。

靠近这栋建造于自然形成的高地之上的房屋，未来居住于这里的住户可以360° 无死角地欣赏到这里令人叹为观止的原野景观，面对这些，人们会禁不住怀疑是否能将其定义为一栋房屋，还是干脆赋予它一个新的名词。

独奏屋的设计看似提供了所有的基本结构要素：混凝土地基（被设计成一个圆形的"通道"）、与三个居住区呼应的立面（采用排孔设计并在需要时可以沿着环形外墙滑动）、一个简单的平屋顶（上面包括所有的技术性基础设施——太阳能光伏板、储水箱和发电机——全部裸露放置于屋顶，似雕刻品一般）以及一整套家具（全部直接与提供支撑力的立柱相连）。在该项目中，所有的传统都受到了质疑，随后又都得到了可靠的新答案。

太阳能光伏板将提供热能和电能，这些能量将被储存在缓冲罐中。用水将就地取材，并在净化后循环使用。每一种设备都像抽象物品一样放置于屋顶。

该建筑的面积为1600m²，包括一处面积为1050m²的内院（带有一处天然泳池）以及三个分区——起居室、主卧室和客房（每个分区的面积均为60m²）。这座房屋的设计实现了既独立又壮观、既正式又稀疏、既开放又内向、既透明又不透明、既豪华又简朴的效果。

Solo House

The Solo Houses project in Matarraña, south of Barcelona was initiated by French developer Christian Bourdais. This contemporary small resort-prototypes are conceived by young international architects.

The first house was completed three years ago by the Chilean architects duo Mauricio Pezo and Sofia Von Ellrichshausen, and the second house by the Belgium architects Office KGDVS was inaugurated on April 24th.

The second Solo house frames the top of a plateau, overlooking the surrounding forests and dramatical landscape. Since the scenery is so impressive, the Office KGDVS felt architecture should be invisible, merely emphasising the natural qualities of the surroundings. A simple circular roof with a diameter of 45 meters underlines the qualities of both the plateau and its edge. The roof functions as a shelter, and forms the perim-

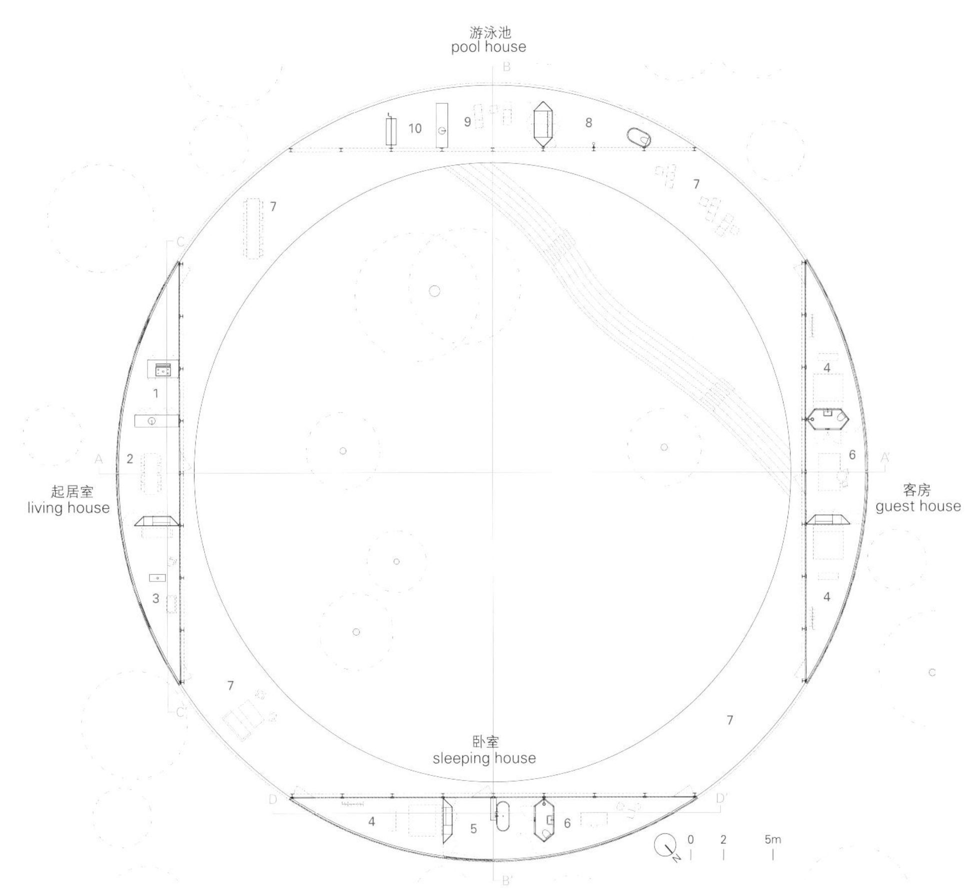

1. 厨房 2. 餐厅 3. 起居室 4. 卧室 5. 浴室 6. 书房 7. 露台 8. 技术设备间 9. 日光浴室 10. 室外厨房
1. kitchen 2. dining 3. living 4. bedroom 5. bathroom 6. study
7. terrace 8. technical 9. solarium 10. exterior kitchen

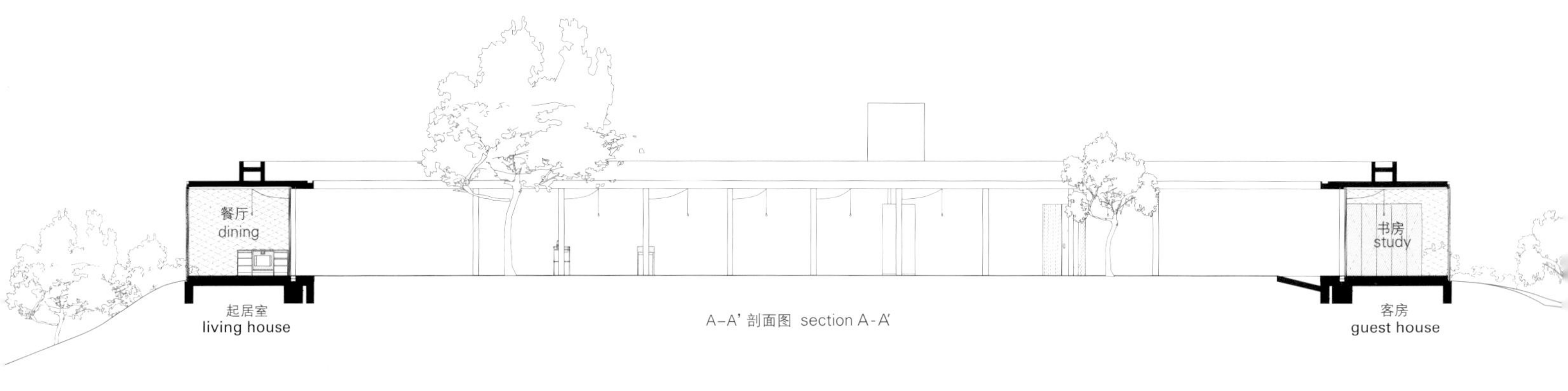

A–A' 剖面图 section A-A'

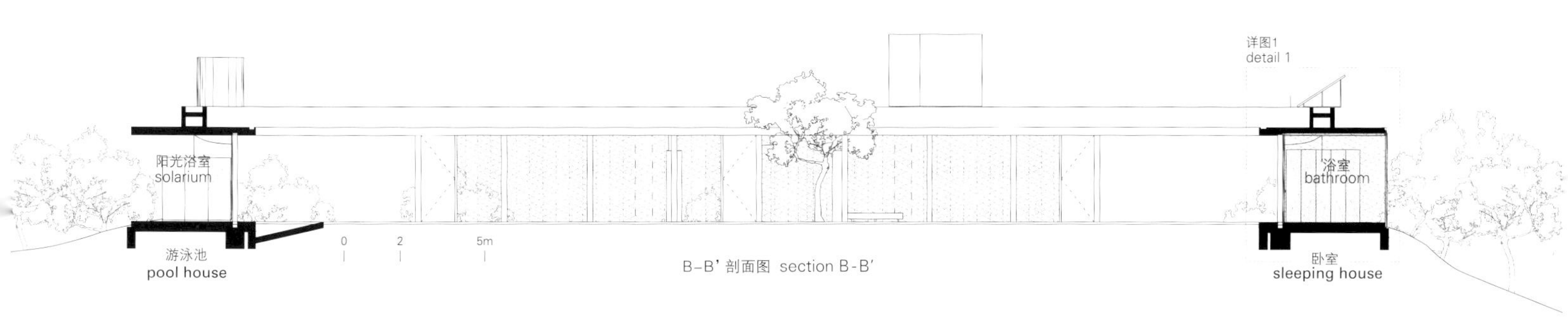

B-B' 剖面图 section B-B'

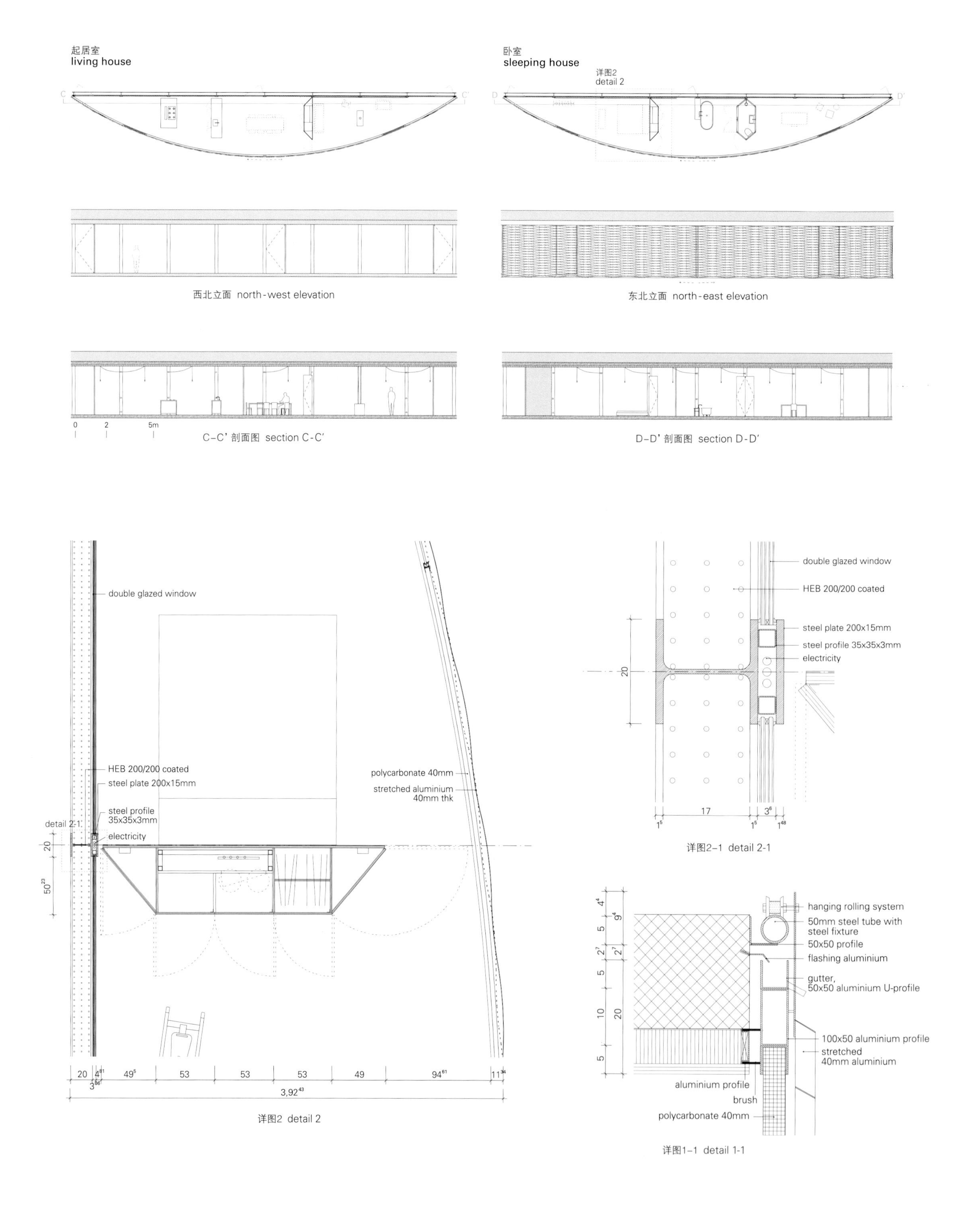
起居室
living house
卧室
sleeping house
详图2
detail 2
西北立面 north-west elevation
东北立面 north-east elevation
C-C' 剖面图 section C-C'
D-D' 剖面图 section D-D'
double glazed window
HEB 200/200 coated
steel plate 200x15mm
steel profile 35x35x3mm
electricity
detail 2-1
polycarbonate 40mm
stretched aluminium 40mm thk
详图2 detail 2
steel profile 35x35x3mm
详图2-1 detail 2-1
hanging rolling system
50mm steel tube with steel fixture
50x50 profile
flashing aluminium
gutter,
50x50 aluminium U-profile
100x50 aluminium profile
stretched 40mm aluminium
aluminium profile
brush
polycarbonate 40mm
详图1-1 detail 1-1

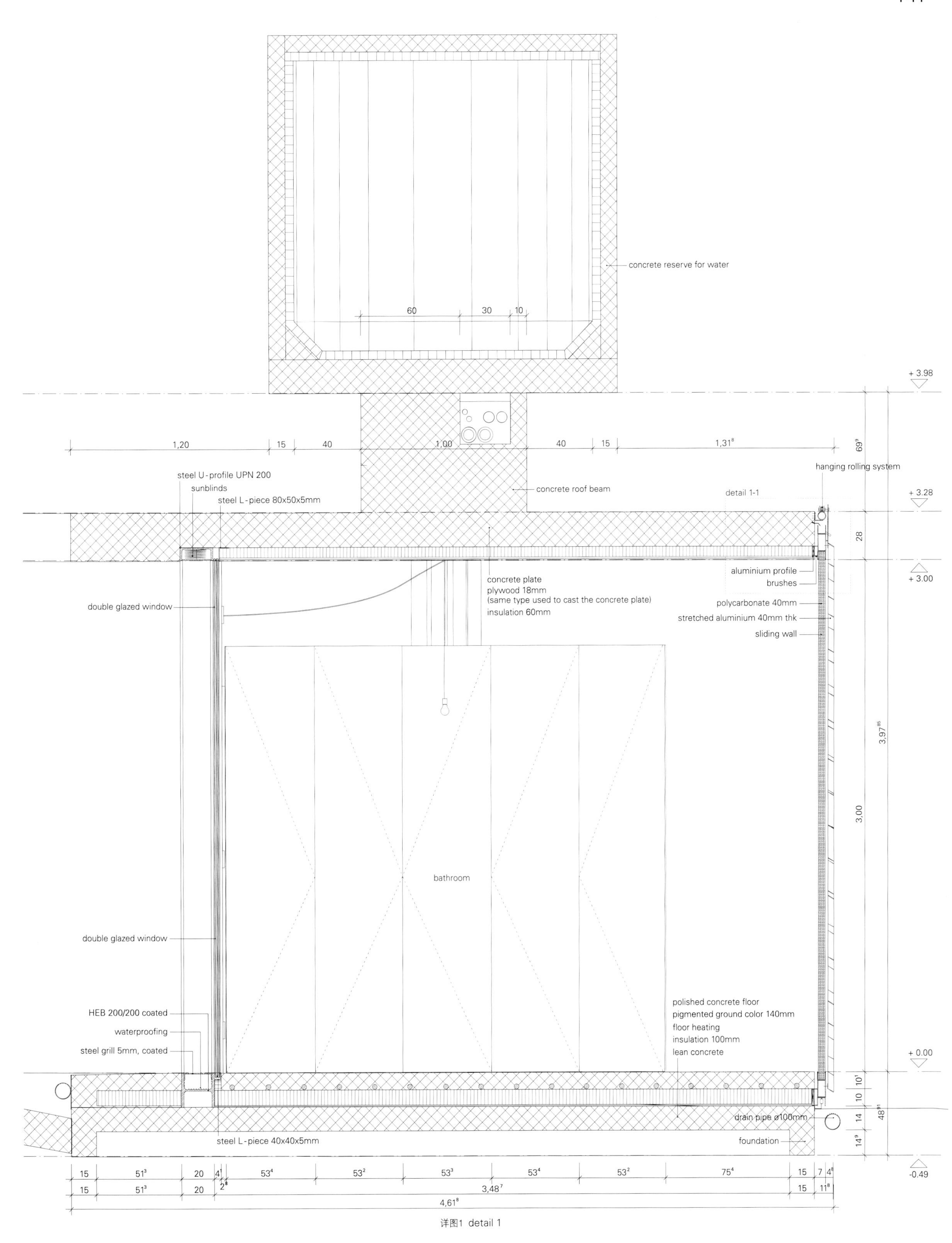
concrete reserve for water
60
30
10
+ 3.98
1,20
15
40
1,00
40
15
1,31
69
hanging rolling system
steel U-profile UPN 200
sunblinds
steel L-piece 80x50x5mm
concrete roof beam
detail 1-1
+ 3.28
28
aluminium profile
brushes
+ 3.00
concrete plate
plywood 18mm
(same type used to cast the concrete plate)
insulation 60mm
double glazed window
polycarbonate 40mm
stretched aluminium 40mm thk
sliding wall
3,97
3,00
bathroom
double glazed window
HEB 200/200 coated
waterproofing
steel grill 5mm, coated
polished concrete floor
pigmented ground color 140mm
floor heating
insulation 100mm
lean concrete
+ 0.00
drain pipe ø100mm
foundation
steel L-piece 40x40x5mm
-0.49
15
51
20
53
75
3,48
4,61

详图1 detail 1

eter of the inhabited surface. It is supported by four straight rows of nine columns, which cut chords from the circular base shape. Only these four areas are inhabited, with variable levels of protection. Large stretches of curtain facade slide on the outer edge of the circle, allowing the living areas to fully open, and providing a maximum relationship between the dweller and the surrounding nature. Since the terrain is not connected to any services, the house will be completely self-dependent.

Approaching the house that is built on a natural plateau, providing its future residents a breath-taking 360º panorama of wilderness, one immediately wonders whether this can be put into the definition of a house at all or whether it needs a new term altogether.

The house seemingly provides all the basic structural elements: a concrete foundation (but shaped as a circular "catwalk"), a façade corresponding with the three inhabited segments (but perforated and with the feature to be shifted around the circumference if needed), a simple flat roof (but with all technical infrastructure - photovoltaic panels, water tanks and generators - exposed as though sculptural objects), and a set of furniture (but all directly connected to the supporting columns). In this project all conventions are questioned and then given a solid new answer.

Photovoltaic panels will provide thermic and electrical energy, which will be stocked in buffer tanks. Water is claimed on the site itself, and purified after use. Each of the devices are placed as abstract objects on the roof.

The overall 1,600m² surface includes a 1050m² patio garden with a natural carved pool and three segments – living room, master bedroom and guest room, each 60m². It is at once discrete and imposing, ceremonial and sparse, open and introvert, transparent and opaque, luxurious and austere.

项目名称：Solo House / 地点：Polygon 13, Parcel 381, Cretas, Province Teruel, Spain
建筑师：OFFICE Kersten Geers David Van Severen
参与设计建筑师：Diogo Porto Architects
合作方：Jan Lenaerts / 艺术设计：Pieter Vermeersch
承包商：Constructions Ferras Prat, sl
结构工程师：Laguens Arquitectos Asociados, sl
技术建筑师：Alfonso Arrufat
装置项目：Alex Montañes-Montañes Instalaciones
立面：De Clercq / 客户：Christian Bourdais
用地面积：92,889m² / 建筑面积：550m²
造价：EUR 1,000,000 / 竣工时间：2017
摄影师：©Bas Princen (courtesy of Solo House)

公园中的亭阁建筑

新建的国际摩托车联盟（FIM）总部位于连接日内瓦和沃州的铁路和州际公路之间，项目所在的坡地上长满树木，使得新总部仿佛一栋位于公园中的建筑。该建筑位于自然景观优美的基地地势较低的地方，在附近看去，其圆形的结构给人留下了深刻的印象。

加速度、速度和动力学

建筑设置在从地面升起的基座之上，上方起防护作用的宽阔平屋顶由精巧的柱子支撑，它在多样化的建筑环境中显得卓尔不群，成了人们关注的焦点。建筑的主体为圆形，而由森林般的柱网连接的椭圆形楼板交错布置，让人联想到摩托车竞技运动的速度和动感。当人们驾车从州际公路或乘火车在铁道上经过时，垂直柱网的节奏感和立面的厚度将为人们呈现一种动态的效果。

人们可以通过项目场地旁边的一条小径进入建筑。主入口与入口通道直接相连，而位于建筑北侧的次要入口连接员工停车场。

光与透明性

FIM新总部取代了先前被拆除的总部建筑。新总部由地上两层和地下室组成，一层设有两个入口，与立面垂直。这两个入口定义了规则的平面布局网格，将使用者引入了中央大厅，在那里他们可以进入不同的功能区。建筑一层容纳主要的公共空间：东侧为礼堂和训练室，南侧为自助餐厅和展览空间。空间采用模块化设计，实现了功能的灵活性。

建筑室内的中央有一段宏伟的旋转楼梯，连接两个楼层，自然光通过上部的圆顶天窗照亮整个空间。楼梯的螺旋形式延续了入口大厅向上移动的态势，将人们引领至上层的行政和管理设施中。作为一个独立的单元，楼梯由混凝土浇筑而成，其下部呈三角形，如同脊椎结构一般支撑着建筑整体的透明框架。

用户舒适度

建筑的技术设施最大化地确保了用户使用的灵活性。在外围的办公区，热敏楼板系统通过天花板提供供暖和制冷，而通风系统和电网

FIM 新总部

Localarchitecture

则安装于活动地板之下。除了一层的地暖设施之外，大厅以及交通流线区域都没有安装任何技术设施。办公室天花板上安置的圆形挡板能够控制建筑的隔声效果。一层的外部遮阳系统和上层伸出的屋顶板是根据当地太阳运行的轨迹而设计的，能够有效地控制由于光照造成的季节性过热。

New FIM Headquarters

A pavilion in a park

Set between the railway and the cantonal road connecting Geneva to the canton of Vaud, on a sloping terrain with trees, the new international headquarters of the motorcycling world has the air of a pavilion in a park. The building occupies the lower part of the naturally landscaped plot, an imposing circular presence when seen from the adjacent roundabout.

Acceleration, speed and kinetics

Set on a base which raises it above ground level and protected by a wide flat roof supported by fine columns, the building stands out as the focal point in a diverse architectural context. Its circular forms evoke the movement and speed of the motorcycling world, suggested by the dynamic arrangement of the offset oval slabs connected by a forest of pillars. The vertical rhythm of the pillars and the depth of the façade produce a kinetic effect when viewed by passing drivers on the cantonal road or passengers on the railway.

The building is accessed by a path adjacent to the site. The main entrance connects directly to the access road while a secondary entrance on the north side of the building connects to the staff car park.

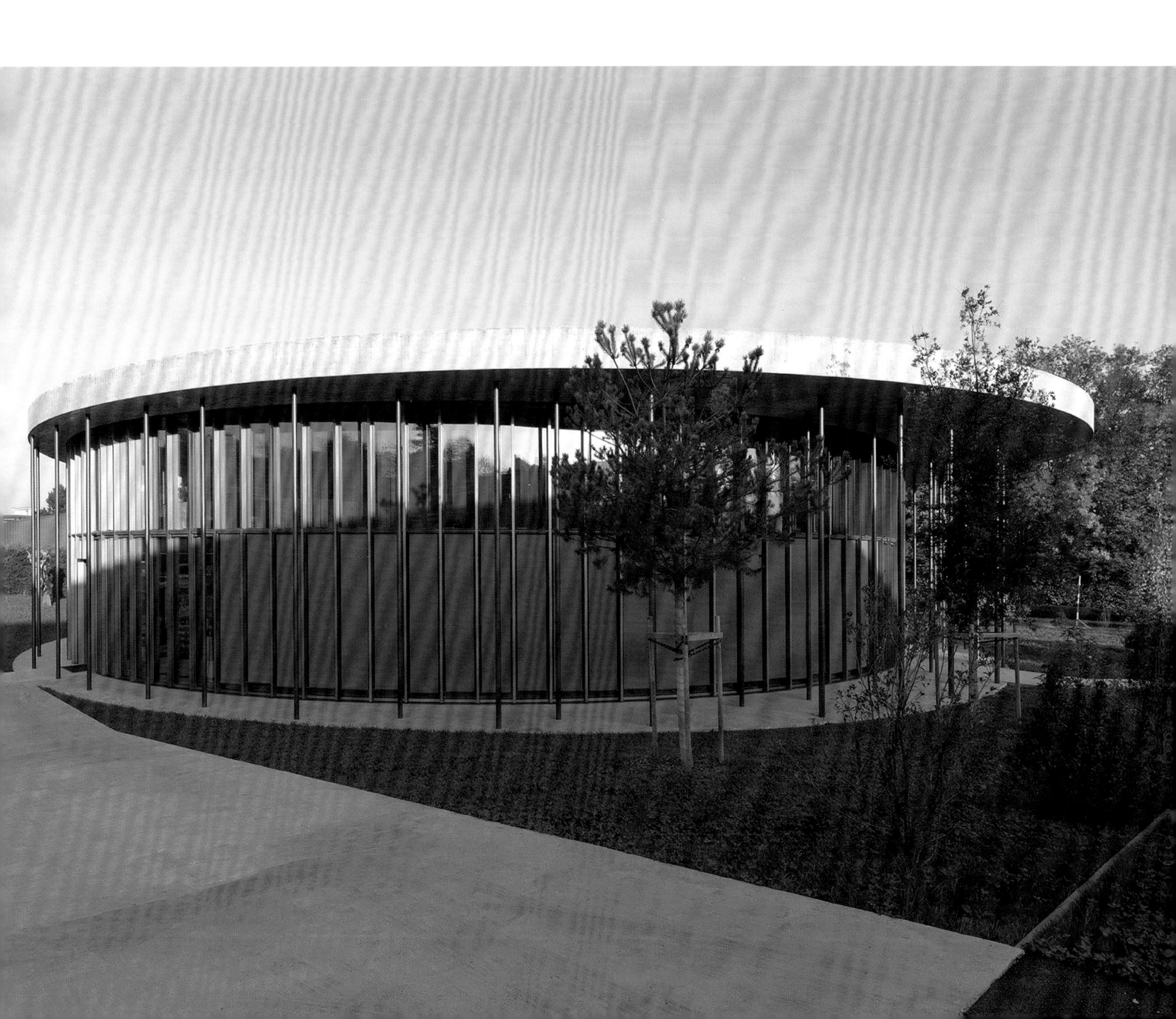

Light and transparency

The new FIM building replaces the former headquarters, which was demolished. It comprises two storeys over the existing basement level and is accessed by two entrances, perpendicular to the façade, on the ground floor. They define the regular grid of the floorplan, leading users to a central hall which provides access to the various functions. The ground floor houses the major communal spaces: the auditorium and the training room on the east side, the cafeteria and exhibition space to the south. The spaces are designed to be flexible and modular.

At the heart of the building, with natural lighting from the skylight domes, is a monumental staircase that connects the two levels. Its spiral form extends the upward movement of the entrance hall, leading towards the administration and management facilities on the upper storey. Cast in concrete

as a single unit, its triangular underside suggests a vertebrate structure – like a spinal column bearing the transparent framework of the building as a whole.

User comfort

The building's technical facilities were developed to ensure maximum flexibility for its users. In the peripheral office areas, the thermally active slab system provides heating and cooling from the ceiling, while the ventilation system and electricity network are fitted below the raised floor. The hall and circulation areas are free of all technical installations except for the floor at ground level, which is heated. Building acoustics are managed via circular baffles arranged on the office ceilings. Seasonal overheating from solar energy is managed at ground level by a system of external blinds and on the upper storey by the oversized roof slab, whose contour is designed to match the sun's pathway across the sky.

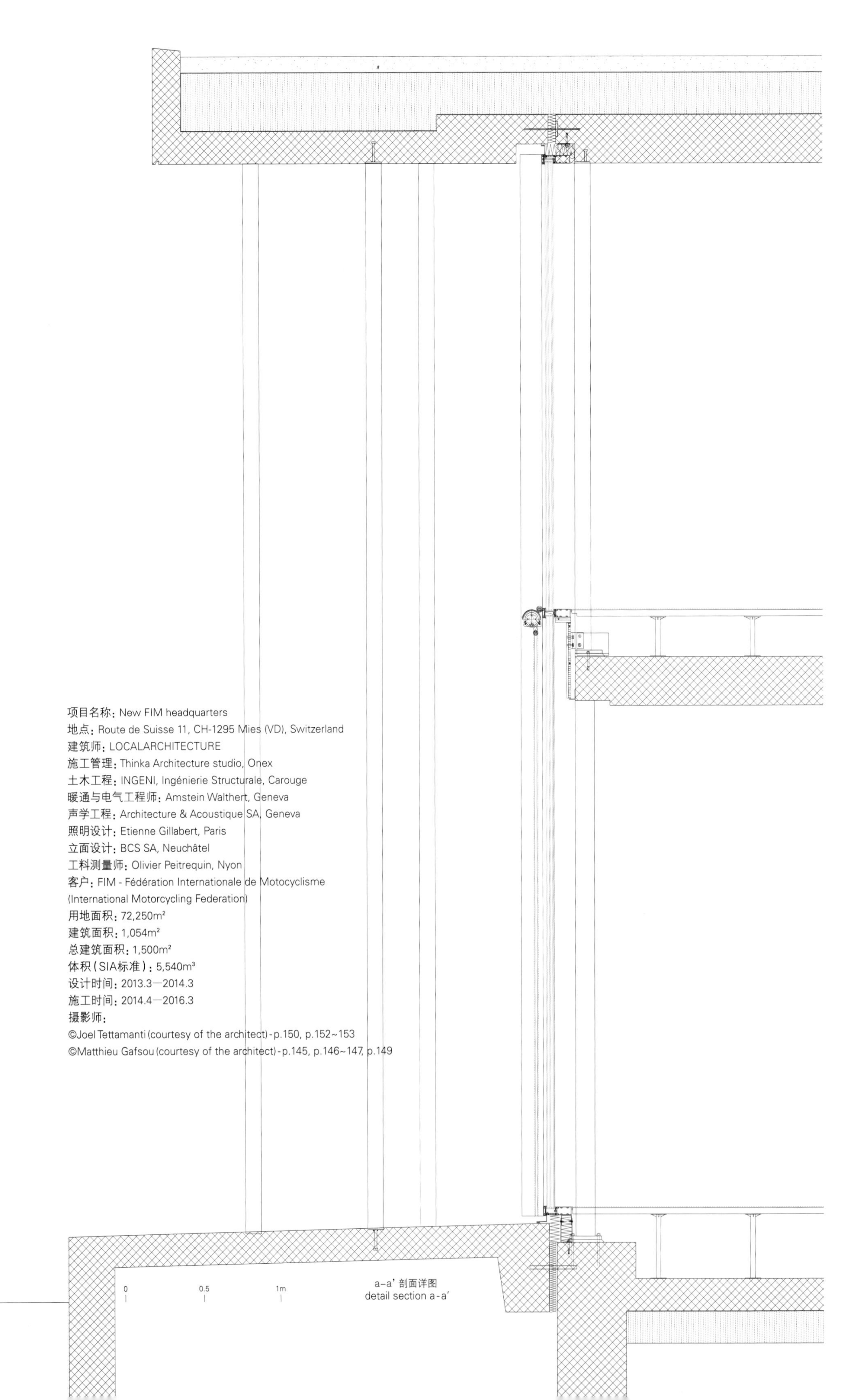

a–a' 剖面详图
detail section a-a'

项目名称：New FIM headquarters
地点：Route de Suisse 11, CH-1295 Mies (VD), Switzerland
建筑师：LOCALARCHITECTURE
施工管理：Thinka Architecture studio, Onex
土木工程：INGENI, Ingénierie Structurale, Carouge
暖通与电气工程师：Amstein Walthert, Geneva
声学工程：Architecture & Acoustique SA, Geneva
照明设计：Etienne Gillabert, Paris
立面设计：BCS SA, Neuchâtel
工料测量师：Olivier Peitrequin, Nyon
客户：FIM - Fédération Internationale de Motocyclisme (International Motorcycling Federation)
用地面积：72,250m²
建筑面积：1,054m²
总建筑面积：1,500m²
体积（SIA标准）：5,540m³
设计时间：2013.3—2014.3
施工时间：2014.4—2016.3
摄影师：
©Joel Tettamanti (courtesy of the architect) - p.150, p.152~153
©Matthieu Gafsou (courtesy of the architect) - p.145, p.146~147, p.149

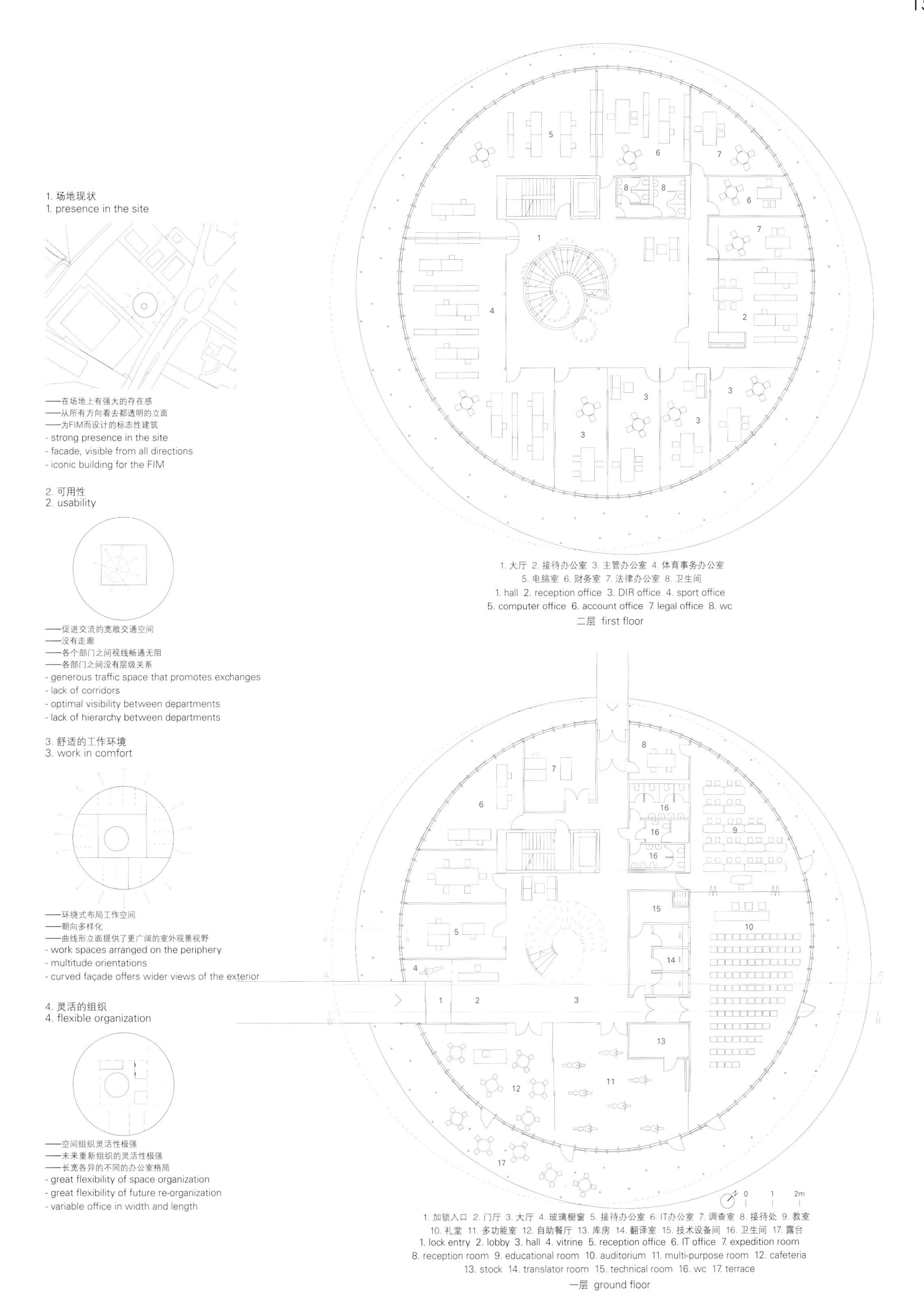

1. 大厅 2. 接待办公室 3. 主管办公室 4. 体育事务办公室 5. 电脑室 6. 财务室 7. 法律办公室 8. 卫生间

1. hall 2. reception office 3. DIR office 4. sport office 5. computer office 6. account office 7. legal office 8. wc

二层 first floor

1. 加锁入口 2. 门厅 3. 大厅 4. 玻璃橱窗 5. 接待办公室 6. IT办公室 7. 调查室 8. 接待处 9. 教室 10. 礼堂 11. 多功能室 12. 自助餐厅 13. 库房 14. 翻译室 15. 技术设备间 16. 卫生间 17. 露台

1. lock entry 2. lobby 3. hall 4. vitrine 5. reception office 6. IT office 7. expedition room 8. reception room 9. educational room 10. auditorium 11. multi-purpose room 12. cafeteria 13. stock 14. translator room 15. technical room 16. wc 17. terrace

一层 ground floor

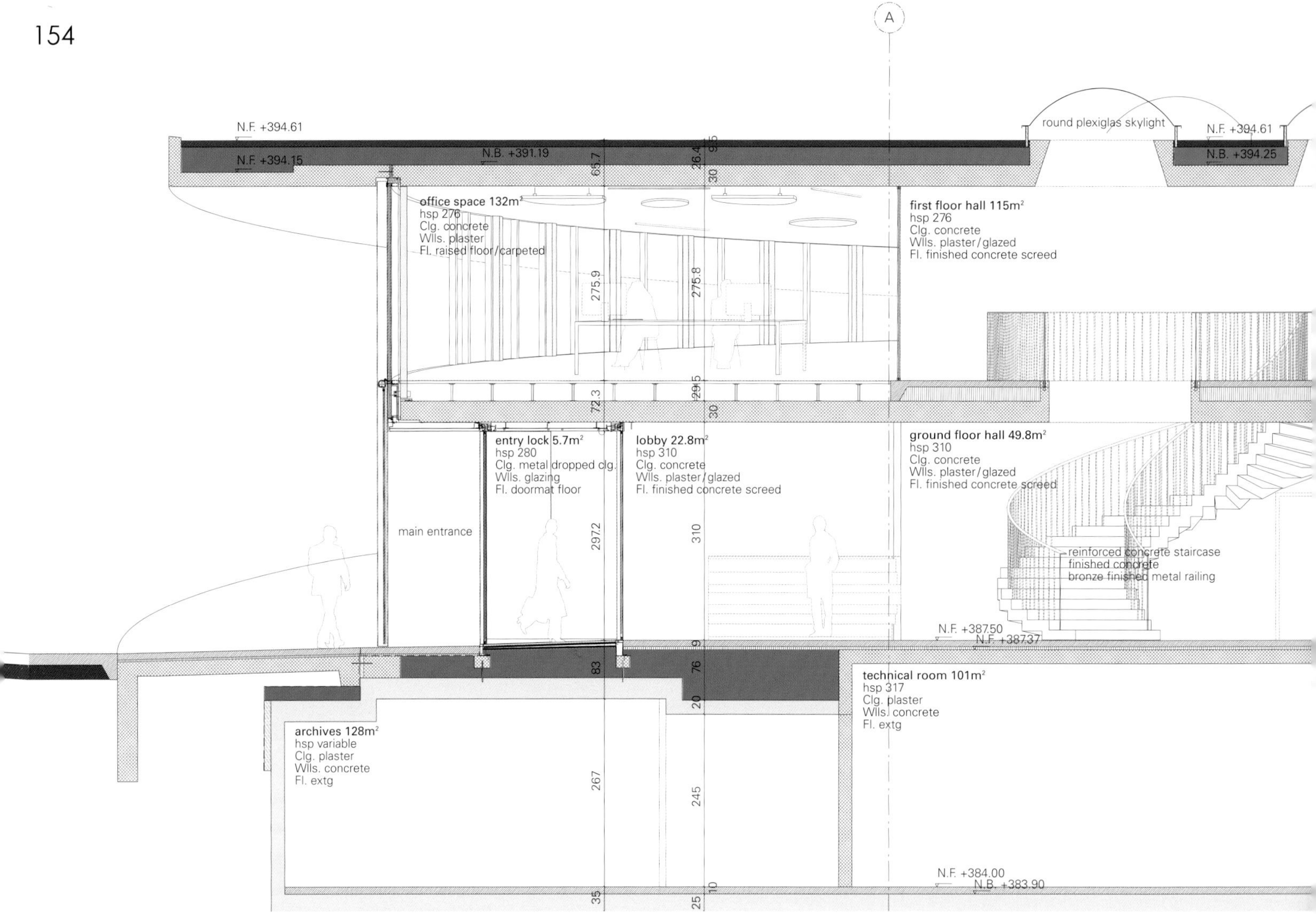
A
round plexiglas skylight
N.F. +394.61
N.F. +394.15
N.B. +391.19
N.F. +394.61
N.B. +394.25
office space 132m²
hsp 276
Clg. concrete
Wlls. plaster
Fl. raised floor/carpeted
first floor hall 115m²
hsp 276
Clg. concrete
Wlls. plaster/glazed
Fl. finished concrete screed
entry lock 5.7m²
hsp 280
Clg. metal dropped clg.
Wlls. glazing
Fl. doormat floor
lobby 22.8m²
hsp 310
Clg. concrete
Wlls. plaster/glazed
Fl. finished concrete screed
ground floor hall 49.8m²
hsp 310
Clg. concrete
Wlls. plaster/glazed
Fl. finished concrete screed
main entrance
reinforced concrete staircase
finished concrete
bronze finished metal railing
N.F. +387.50
N.F. +387.37
technical room 101m²
hsp 317
Clg. plaster
Wlls. concrete
Fl. extg
archives 128m²
hsp variable
Clg. plaster
Wlls. concrete
Fl. extg
N.F. +384.00
N.B. +383.90
65.7
76.4
275.9
275.8
72.3
297.2
310
83
76
20
267
245
35
25

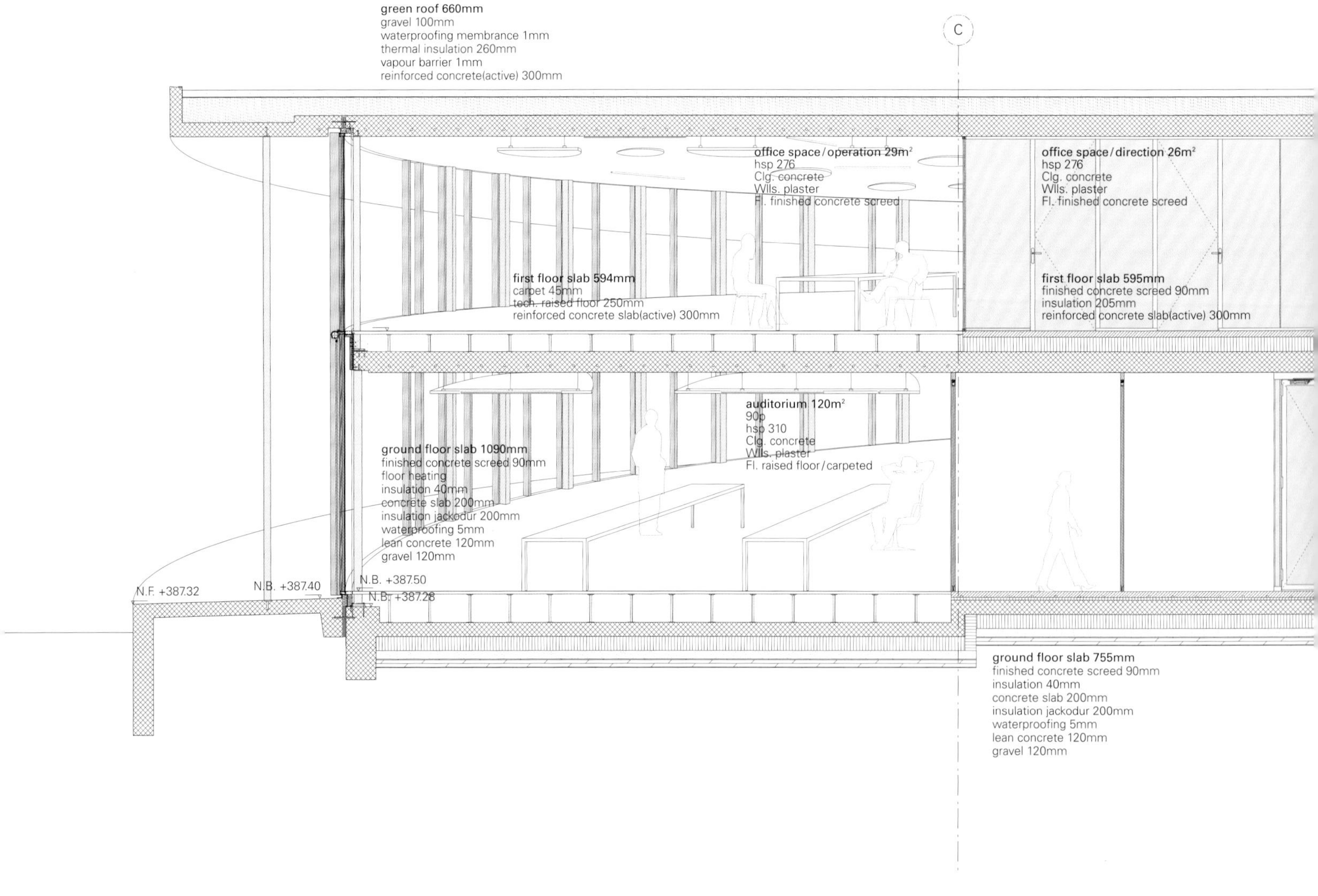
green roof 660mm
gravel 100mm
waterproofing membrance 1mm
thermal insulation 260mm
vapour barrier 1mm
reinforced concrete(active) 300mm
C
office space/operation 29m²
hsp 276
Clg. concrete
Wlls. plaster
Fl. finished concrete screed
office space/direction 26m²
hsp 276
Clg. concrete
Wlls. plaster
Fl. finished concrete screed
first floor slab 594mm
carpet 45mm
tech. raised floor 250mm
reinforced concrete slab(active) 300mm
first floor slab 595mm
finished concrete screed 90mm
insulation 205mm
reinforced concrete slab(active) 300mm
auditorium 120m²
90p
hsp 310
Clg. concrete
Wlls. plaster
Fl. raised floor/carpeted
ground floor slab 1090mm
finished concrete screed 90mm
floor heating
insulation 40mm
concrete slab 200mm
insulation jackodur 200mm
waterproofing 5mm
lean concrete 120mm
gravel 120mm
N.F. +387.32
N.B. +387.40
N.B. +387.50
N.B. +387.28
ground floor slab 755mm
finished concrete screed 90mm
insulation 40mm
concrete slab 200mm
insulation jackodur 200mm
waterproofing 5mm
lean concrete 120mm
gravel 120mm

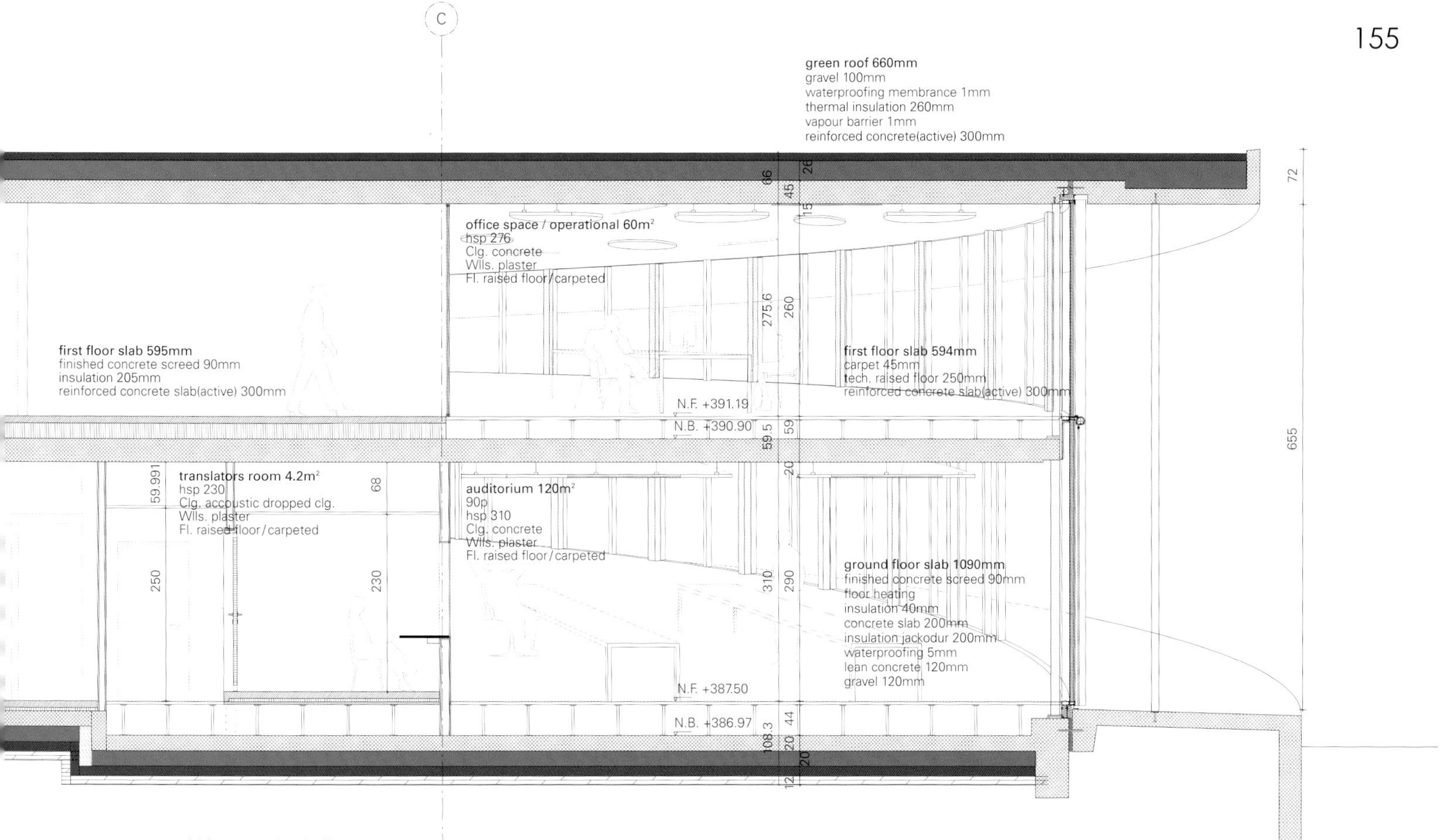
C
green roof 660mm
gravel 100mm
waterproofing membrance 1mm
thermal insulation 260mm
vapour barrier 1mm
reinforced concrete(active) 300mm
office space / operational 60m²
hsp 276
Clg. concrete
Wlls. plaster
Fl. raised floor/carpeted
first floor slab 595mm
finished concrete screed 90mm
insulation 205mm
reinforced concrete slab(active) 300mm
first floor slab 594mm
carpet 45mm
tech. raised floor 250mm
reinforced concrete slab(active) 300mm
N.F. +391.19
N.B. +390.90
translators room 4.2m²
hsp 230
Clg. accoustic dropped clg.
Wlls. plaster
Fl. raised floor/carpeted
auditorium 120m²
90p
hsp 310
Clg. concrete
Wlls. plaster
Fl. raised floor/carpeted
ground floor slab 1090mm
finished concrete screed 90mm
floor heating
insulation 40mm
concrete slab 200mm
insulation jackodur 200mm
waterproofing 5mm
lean concrete 120mm
gravel 120mm
N.F. +387.50
N.B. +386.97
66
45
26
15
275.6
260
72
655
59.5
59
20
59.991
68
250
230
310
290
108.3
44
20
20
12

A–A' 剖面图 section A-A'

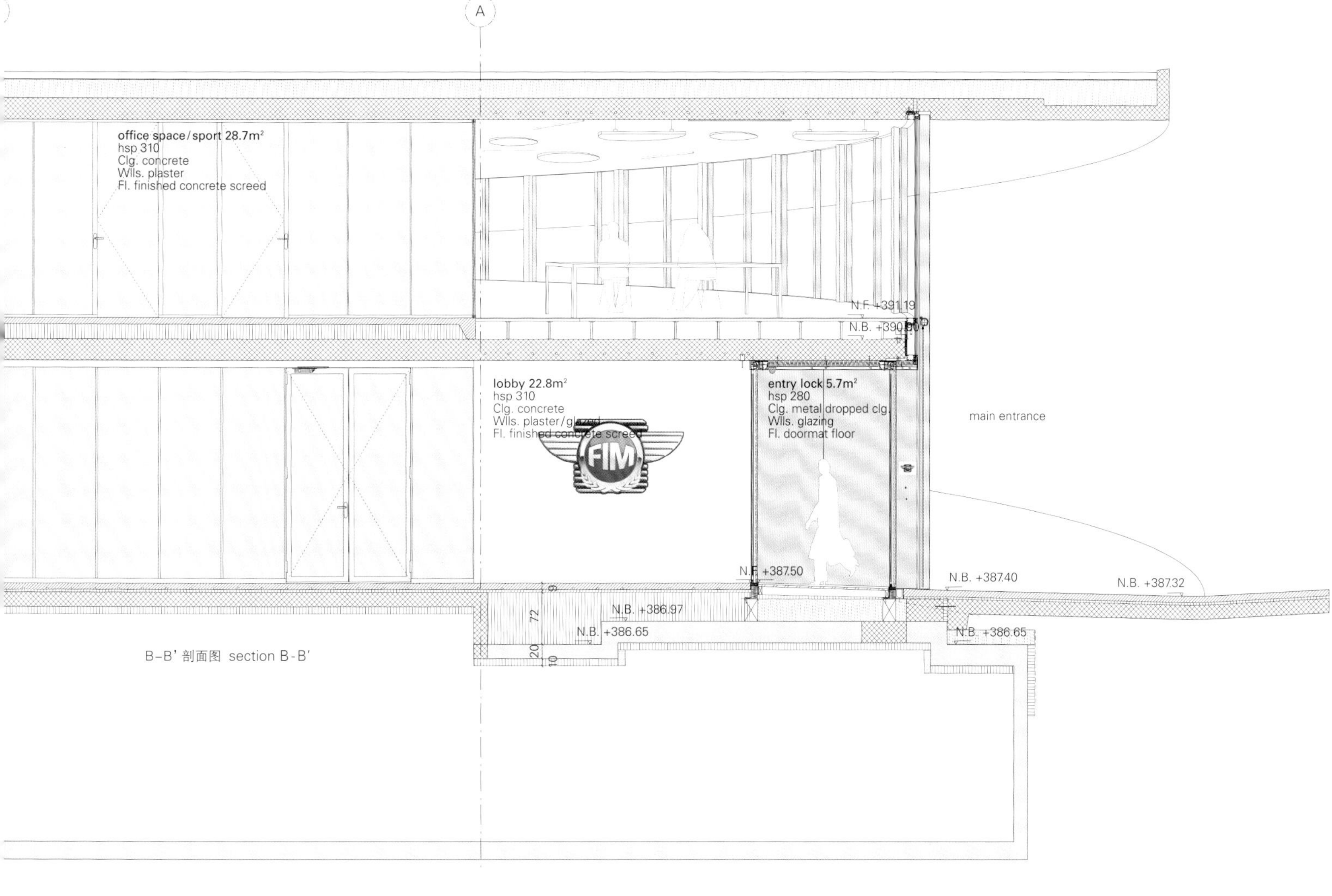
A
office space / sport 28.7m²
hsp 310
Clg. concrete
Wlls. plaster
Fl. finished concrete screed
N.F. +391.19
N.B. +390.90
lobby 22.8m²
hsp 310
Clg. concrete
Wlls. plaster/glazed
Fl. finished concrete screed
entry lock 5.7m²
hsp 280
Clg. metal dropped clg.
Wlls. glazing
Fl. doormat floor
main entrance
N.F. +387.50
N.B. +387.40
N.B. +387.32
N.B. +386.97
N.B. +386.65
N.B. +386.65
FIM
9
72
20
10

B–B' 剖面图 section B-B'

回收和再利用

Recycle and

可持续性发展的两大支柱
The Two Pillars of Sustainable Development

"……我担心，除非我们向只能引发反抗的道德推理敞开大门，否则将不会有下一个世纪的出现。"马克斯·弗里施在1981年就警告世人要提防消费主义文化意识形态的趋势。36年之后的今天，全球变暖迫使人们更加关注生态意识方面的选择，甚至在建筑学领域也不例外，包括作为任何一个现代社会的物质组成部分的先进设计流程。因此，像"再利用"或"回收利用"这样的理念目前普遍存在于全球文化，并且人们正在检验这些理念对社会、经济政策以及世界体系产生的积极影响。这些理念反对超出全球可持续水平之外的资源消耗，促进所有国家体制具备应对变化的适应性，更被称为"弹性"的同义词。可适性再利用、城市再生[1]、创新型回收艺术以及突破传统的实践都是"重获新生"工具箱的升级改造策略，内部和外部空间都与建筑材料密切相关。下面这篇文章将介绍一些再利用剩余产品（如木料）和拆毁建筑垃圾（如石料）的项目。正如下面这篇文章所解释的那样，建筑师和"文化创意者"正在逐渐留意到回收和再利用结合智能设计（例如"从摇篮到摇篮（可持续发展）"方法）所提供的广泛机遇。

"[...] I fear that unless we open up towards moral reasoning which can only arise out of resistance, there will be no next century." Written in 1981, Max Frisch's admonition warned against the trends of the culture-ideology of consumerism. Thirty-six years later, global warming imposes eco-conscious choices, even in architecture, involving cutting-edge processes as a physical part of any modern society. Therefore, concepts such as "Reuse" or "Recycle" are now widespread in global cultures and examined for their positive impact on community, economic policies and the world-system. They fight against the consumption of resources beyond globally sustainable levels, contributing to the adaption of any country-system to cope with changes, better known as a synonym for "Resilience". Adaptive reuse, urban re-activation[1], creative recycled art and non-conventional practices are upcycling strategies for a "second life" toolbox, strictly connected to building materials in both inner and outer spaces. The following essay concerns the presentation of projects that range from the reuse of leftover products (e.g. wood) to demolition waste (e.g. stone). Architects and "cultural creatives" are taking notice of the wide range of opportunities offered by recycling and reuse coupled with smart design (e.g. "cradle-to-cradle" approach), as explained in the article below.

Praz-de-Fort谷仓改造项目_Conversion, Praz-de-Fort/Savioz Fabrizzi Architectes
兰科湖畔的谷仓住宅_Barn House at Lake Ranco/Estudio Valdés Arquitectos
谷仓改造 _Re [Barn]/Circa Morris-Nunn Architects
Kyodo住宅_Kyodo House/SANDWICH + Team Low-energy

回收和再利用：可持续性发展的两大支柱
Recycle and Reuse: The Two Pillars of Sustainable Development/Fabrizio Aimar

无论是西方国家还是东方国家，所有的现代社会都依赖于对可再生或不可再生自然资源的开采。我们的社区每天都会参与到"自然资源经济"的进程当中，致力于全球自然资源的供应、需求和分配。这一全球生态模型的核心目的之一在于能够更好地了解它在经济政策中的地位，以便开发一个有助于可持续增长的有效管理框架。根据这些目标，需要记住的另一个要点是，影响主要是由两个相互联系的原因造成的：人口和每个人的累积倍数效应。许多论证都支持这一观点。首先，尽管资源理念呈现出多样性，并随着新技术的出现而发生变化，但是应该说在一个资源有限的星球上，我们必须接受可持续性。其次，为了减少倾倒至垃圾填埋场的垃圾数量，以及一个国家的整体二氧化碳排放量，废物处理政策层级似乎应当是我们塑造生态友好型生活所必需的共同准则。最后，我们必须记住一点——废物实际上象征着以商品和能源为形式的资源的巨大损失。举例来说，2014年期间欧盟（28个国家）产生了近26亿吨的废物，而仅有一半的废物得到了处理和重新利用（包括回填和能量回收）[2]。鉴于上述提到的事项，亚历杭德罗·阿拉维纳呼吁在意大利举行的主题为"前线报道"的第15届威尼斯国际建筑展览会期间，要加强政策力度和共同作用。值得一提的是，智利的普利兹克建筑奖对2015年威尼斯艺术双年展期间产生的90吨废物进行重新利用，建造了两间活动介绍室。超过11km的铝型材悬挂于天花板下方，搭配10 000m²的分层灰泥墙，在灰泥墙中结合了模块化的陈列方式。但其设计目的是什么呢？设计这种具有煽动性的装置，其目的不仅在于鼓励回收利用、反复利用和升级利用的价值观，同时也能引发人们对于日常行为产生的大量废旧物资的自我反思。

这一理念似乎很好地介绍了这个主题之下的首个项目，在该项目中，建筑师重新利用了来自地震灾区干草棚的大量桁架，组装成

All modern societies, in both western and eastern nations, depend on extracting natural resources, renewable and non-renewable. Our communities are daily involved in Natural Resource Economics processes, committed to the supplying, demanding, and allocating of global natural resources. One of the core goals of this planetary ecological model is to better understand their position in economic policies, in order to develop an effective management framework for sustainable growth. According to these aims, another key point to remember is that the impact is principally due to the product of two interlinking causes: population and multiplied times effect per person. There are many arguments in support of this view. Firstly, despite the fact that the concept of resource is multifarious and modifies with the advent of any new technology, it can be argued that, on a finite planet, sustainability is not an option. Secondly, the waste hierarchy seems a common guideline deemed necessary for shaping an eco-friendly life, in order to reduce the amount of garbage that goes to landfill and, consequently, the overall carbon dioxide emissions in a country. Thirdly, one has to bear in mind that the waste actually symbolises an enormous loss of resources in the form of both goods and energy. To give an illustration of what this means, during 2014 the European Union (28 countries) had produced 2,598,140,000 tonnes of waste, whereas only half was treated and recovered (including backfilling and energy recovery)[2]. According to the aforementioned considerations, Alejandro Aravena made an appeal for stronger policy efforts and common actions during the 15th International Architecture Exhibition in Venice (Italy), titled "Reporting from the Front". In particular, the Chilean Pritzker Price reused 90 tonnes of waste generated by the 2015 Venice Art Biennale to create two introducing rooms for the event. Over

Galpon Ranco (由Estudio Valdés建筑师事务所建造的、位于智利兰科湖边的一处村舍) 的人字形屋顶，该住宅还重新利用了先前谷仓废弃的木料。木质覆层赋予所有房间一种舒适的森林小屋般的感觉，其受到磨损的外观让人回想起了古老的乡土风情。外表的古旧暗指建筑师在建筑现状的分析过程中所体现的克服当代享乐主义思想的理性诚实。事实上，结合了敏锐而又诗意的写实主义，该项目的实施体现了一种与见证和回忆紧密相连的态度。它看起来就像是一本现代的"羊皮书"，也就是一份最初的文字被抹去的手稿，然后将兽皮或羊皮纸重新利用在其他地方。这是一种普遍的趋势，特别是在中世纪的欧洲，通过清洗或刮掉笔迹的方式擦掉之前的文字，以便在纸张上再次进行书写。通过重新利用羊皮纸来制作羊皮书的原因似乎主要在于这么做要比制作新羊皮纸更便宜。在建筑学中，它指的是针对一种目的制造 (锻造、雕刻或手工编织) 出但随后经过改造重新应用于其他目的的任意一种构件，同时还得承认它并不完美。但在"时间的洗礼"下，错误并不是不受欢迎的结果，它能够引发富有创造性的过程，展现出令人意想不到的构造。与之相似，Savioz Fabrizzi建筑师事务所设计的位于瑞士奥西耶尔Praz-de-Fort村庄的项目，意在通过将新的住房需求与历史遗产相结合的方式来激发人们反思传统建筑。这些瑞士建筑师重新利用了一个始建于19世纪的谷仓的外围护结构，创造了一个新的开放型空间住宅。相比之下，该项目在修复和修补的方法方面与Galpon Ranco的项目不同。首先，不论是在外表还是紧凑的尺寸上，新住宅与最初的干草棚都极为相似，尽管采用了新的玻璃洞口设计也没有影响这一点。与智利的项目相比，由于采用木质覆层，连续性概念战胜了简单的装饰性理念。尽管乍一看人们可能会觉得它们十分相似，但这两个项目在内部木质材料布局设计的更新方面有所不同，前者采用饱经风化侵蚀的覆层，而瑞士项目则是一个现代的敞开式空间的空间组合。

11km of aluminium profiles was suspended from the ceiling, coupled with 10,000sqm of layered plaster walls that incorporated modular displays. But what was his purpose? This provocative installation was designed to not only encourage recycling, reusing, and upcycling values, but also lead to self-reflection on the large amount of scrap materials generated by any human activity.

This idea seems a useful introduction to the first project in this topic, wherein a large number of trusses recovered from a quake-affected hayloft were reused for assembling the gable roof in Galpon Ranco, a cottage by Estudio Valdés Arquitectos built along the edge of Lake Ranco in Chile, salvaging the wooden ruins from the former barn. The wood cladding gives all the rooms a cosy, cabin-like feel that recalls, with its worn look, an old rustic charm. The old aspect alludes to an intellectual honesty in the analysis of existing conditions, which overcomes the contemporary hedonistic thought.

In fact, the implementation of this project shows an attitude strictly connected to bearing witness and remembrance, coupled with a sensitive and poetic Verism. It looks like a modern "palimpsest", namely a manuscript on which an earlier text has been erased and the animal-skin paper or parchment reused for another. It was a common tendency, particularly in Europe during the Middle Ages, to rub out an earlier piece of writing by means of washing or scraping the handwriting in order to prepare it for a script once again. The reason for making palimpsests by the economic reuse of parchment seems largely due to the fact that it was cheaper than preparing a new skin. In architecture it indicates any element manufactured (wrought, carved, or hand-woven) for one purpose but subsequently modified and reused in another, while admitting imperfection. Moreover, under "the patina of time", the error is not an undesirable outcome, but triggers a creative process, revealing unexpected configurations. Similarly, Savioz Fabrizzi Architectes' project in Praz-de-Fort, Orsières, Switzerland intends to stimulate a reflection on traditional architecture as a possible way to combine new housing needs with historical legacy. The Swiss architects reuse the envelope of a 19th-century barn to create a new open-space

Praz-de-Fort谷仓改造项目，瑞士
Conversion, Praz-de-Fort, Switzerland

与之一致的是，资源的有限性和明显减少的人类能源消耗也是法国著名经济学家、哲学家和人类学家塞奇·拉脱谢尔的核心主张[3]。其意识形态分析基于对西方世界"进步"和"开发"之间平衡的强烈反思，引发了八个相互依存的原则——减速增长理论的"八个良性循环原则"。其中包括产品的重新利用和废物的回收以及产量的减少和生活消费品的消耗原则，这些原则似乎成为新型"建筑乌托邦"中最令人信服的观点。只要牢记为得出这一观点所使用的方法的局限性（马克思主义者对整个政治制度和西方文化的一种批判），我们就可以将它作为支持其论点的证据。然而，现如今，减少由具体的能源生产过程（用于生产任何商品或服务）所造成的碳足迹已成为全球可持续发展的一个必然选择。正如1998年普利兹克奖得主——加州大学医学院生理学教授贾雷德·戴蒙德[4]研究和公布的那样，历史证据似乎表明了一个明确的状况。他向人们展示了一个不太直观的画面，但同样，一个有影响力的声音在断言，生态危机似乎是社会崩溃过程中的一个突出问题。而且，这并不是支持此种假设唯一的证据。即使是在今天，人们仍然会在经济繁荣时期扩大生产和增加人口，而忘记（或故意认识不到）这种繁荣时期也可能会变得很糟糕，这种趋势是可以理解的。尽管对环境造成破坏被认为是道德败坏的行为，我们仍在推动世界范围内林业产品消耗量的上升。再加上来自文献的证据，关于这方面的统计资料描绘出了这一令人担忧的状况下的令人信服的观点。由农耕引发的对森林的滥砍滥伐对地球造成了巨大的损失，其中超过一半的损失发生于近55年，这也因为自1961年以来全球纸张消耗量增加了5倍之多[5]。尽管电脑得到了普遍应用，但我们使用（和浪费）纸张的数量并未呈减少趋势，尤其在日本更是如此。事实上，2015年，日本仍是全球第三大纸张和硬纸板（占全球7%）消费国，也是第六大锯材（占全球4%）消费国[6]。因此，由SANDWICH + Team Low-energy设计的位于日本东京的Kyodo住宅项目，体现了在力求实现循环经济策略规定的零填埋目标方面做出的实实在在的努力。建筑师将艺术品

house. By contrast, this project differs from the Galpon Ranco in the approach to restoration and repair. First of all, the new cabin is significantly similar to the original hayloft, both in its appearance and compact dimensions, despite the new glazed openings. Compared to the Chilean project, the notion of continuity overcomes the simple idea of decoration attributed to the timber cladding. Although, at first sight, one may consider them similar, these two projects also differ in the renewal of their inner wooden layouts, where the first one has a weather-worn cladding, and the Swiss project is a modern, open-spaced spatial mix.

Coherently, the finite nature of resources and a considerable reduction of anthropogenic energy consumption are also the core statements postulated by the renowned French economist, philosopher and anthropologist, Serge Latouche[3]. His own ideological analysis is based on a dramatic rethink of the "progress" and "development" balance of the western world, captured in eight inter-dependent principles, the "8 Rs" of degrowth. Re-use products and Re-cycle waste are two of them, as well as the Reduction of production and consumption of consumer goods, and seem to be the most compelling points for a new "concrete utopia". We may use this as evidence to support his argument, provided that we bear in mind the limitations of the methods used to obtain it (a Marxist criticism of the political systems and western culture as a whole). However, reducing the carbon footprint caused by the embodied energy processes required to produce any goods or services is, today, considered an inevitable choice for a universal sustainable development. The historical evidence appears to suggest a clear-cut situation, as studied and publicised by the 1998 Pulitzer Prize Winner, Jared Diamond[4], the professor of physiology at UCLA School of Medicine. He presented a somewhat less straightforward picture, but, likewise, an influential voice in affirming that the ecological crisis seemed one of the highlights in a societal collapse. Furthermore, this is not the only evidence that supports this hypothesis. Even today, there is an understandable human tendency to increase manufacturing and population during good times, forgetting (or deliberately not realising)

谷仓改造，澳大利亚
Re [Barn], Australia

味融入木材边角料在外立面上形成的无规则排列方式中，尝试消除基于大规模林业产品进口的不可持续的经济模式，提出了一种具有创造性的方法来重新利用材料。值得注意的是，我们可以将赋予建筑材料二次生命看作是一种巧妙的方法，既可以降低成本，也减少了因其他过程而产生的二氧化碳排放。这种说法看起来十分中肯，因此，要记住设计师古代裕一所说的："……将剩余的木材制成木片的成本超过了它们的售价。"除了生产力悖论，它还引发了一个问题，那就是我们是否能利用生命周期法刺激并支持地方政策向绿色经济的转变。然而，生命周期思想（LCT）是通过生命周期管理（LCM）过程运作的。前面提到的一些潜在的重要问题与建筑业主/买家（或利益相关者）的环境敏感性以及他们所付诸的行动有关。例如，社会责任和利益相关者的参与对于提供设计场景至关重要，比如接下来将要介绍的由Circa Morris-Nunn建筑师事务所设计的位于澳大利亚的谷仓改造项目，该项目采用具体的方法将剩余的旧材料与当代需求结合起来，形成了良好的建筑实践，提供了一个有趣的案例研究。新建的"附属建筑"仿照传统的殖民时期庄园主住宅的风格，在其施工中使用的许多构件都来自于一处始建于19世纪的谷仓，这个旧谷仓的木框架和承重石墙在一场猛烈的暴风雨中坍塌了。即使是在该案例中，重建实践也经历了串联的转换和设计过程，以及一个生物地质化学循环。意大利锡拉丘兹大学的高级讲师温琴佐·拉蒂纳也以一种相似的方式在思考我们周围的世界，他在2015年被意大利国家建筑师委员会授予"意大利最佳建筑师"的称号。他指出，建筑绝对必须采用现代化风格，但同时也在不断地重塑与古典建筑之间的交换关系。这些术语可能会不时地发生变化，而真正的任务在于重新解读现状。因此，它势必会与

that such times could turn into bad ones. Even though the damage that people inflict on their environment is considered morally reprehensible behaviour, we push worldwide consumption of forest products upwards. Coupled with evidence from literature, the statistics paint a compelling view of this alarming situation. More than half of those losses have occurred within the past 55 years, owing to forest clearance for agricultural purposes, and because the global consumption of paper has grown over five-fold since 1961[5]. Notwithstanding the computer usage, the amount of paper we use (and waste) is not decreasing, especially in Japan. In fact, in 2015, the country was still the third major paper and paperboard consumer worldwide (7 per cent) and the sixth concerning sawn wood (4 per cent)[6]. Therefore, the Kyodo House project by SANDWICH + Team Low- energy in Tokyo, Japan, shows a real effort in seeking to achieve the zero-landfill goal, as stated by the Circular Economy strategy. Combining the taste in art with a random arrangement of wooden offcuts on external façades, the architects have tried to counteract the unsustainable economy model based on massive forest products import, proposing a creative way for reuse of the material. Significantly, giving building materials a second life can be seen as a smart way to reduce both costs and CO_2 emissions arising from any other additional processes to transform them. It seems pertinent, therefore, to remember the words spoken by the planner Yuichi Kodai, who said: "[...] making the leftover timbers into wood chip costs more than their sale value." Beyond the productivity paradoxes, it also raises the question as to whether the life cycle approach is used to stimulate and support the transition to a green economy by local politics. However, Life-Cycle Thinking (LCT) is made operational through Life Cycle Management (LCM). Some potentially important issues already mentioned are connected to the context-sensitivity of the building owners/buyers (or stakeholders) and their actions. For instance, social responsibility and stakeholders' engagement are essential to inform scenarios such as the following, [RE] BARN by Circa Morris-Nunn Architects, in Australia, offers an interesting case study due to the good practices in concrete ways to combine the old remains with contemporary needs. New "outbuildings" replicate the traditional main ones of the colonial manor house and a large number of component parts used in

1. Giampiero Venturini and Carlo Venegoni, Re-Ac (Tools for Urban Re-Activation), D Editore, New Generations, Rome 2016
2. European Commission, Policies, information and services, Statistics, Eurostat – European Statistics, Waste statistics, 2016 http://ec.europa.eu/eurostat/statistics-explained/index.php/Waste_statistics
3. Serge Latouche, Petit traité de la décroissance sereine, Mille et Une Nuits, Fayard, Paris, 2007
4. Jared Diamond, Collapse. How Societies Choose to Fail or Succeed, Viking Penguin, New York, 2005
5. Food and Agriculture Organization of the United Nations (FAO), FAOSTAT, Forestry Production and Trade, http://www.fao.org/faostat/en/#data/FO
6. Food and Agriculture Organization of the United Nations (FAO), Forest products statistics, http://www.fao.org/forestry/statistics/80938@180723/en/

对建筑项目状态的重新解读有关，从而提供新的解答，更新输入数据，并最终形成一个新的成果。总之，它看起来是一种组合，根据一栋建筑的使用寿命而包含了背叛、转化和更新，这一周期的结束也就意味着再生过程的开始。另一方面，现如今我们认为的"高保真"隐约地与各种各样的零售商品有关，例如智能手机或电视机屏幕分辨率（1080p、2K、UHD、4K、8K大小的像素等等）。然而，背叛的观念实际上是对过去那些改变自身的事物一个周期性的再造过程，但这种观念仍僵化不变，尤其是在西方的文化认同中。

从整体来看，如果我们想要解决最为紧迫的环境问题，就必须从能源消耗入手。例如，瑞典最近提出一项立法，将自行车维修、服装和鞋类的定期税从25%减到了12%，但同时，这个北欧国家也是以物美价廉的家具和家居饰品而闻名的品牌——宜家的故乡。仅在20年之前，即1996年的一则标题为"赶走你的印花棉布"的商业广告中，宜家向观众展示了英国消费者在骚乱中反抗他们的生活方式并将廉价的家居饰品扔进大废料桶的画面。而鉴于目前我所讨论的这些内容，不变的风险似乎仍然存在：在一个一次性用品横行的世界中进行本体论的宣传。

the construction were recovered from an original nineteenth-century barn, the timber framing and load bearing stone walls of which had collapsed because of a violent thunderstorm. Even in this example, the rebuilt practice goes through a concatenated transformation and processes, as well as a biogeochemical cycle. A similar way of thinking about the world around us was also expressed by Vincenzo Latina, Senior Lecturer at the University of Syracuse (in Italy) and awarded with the title of "the best Italian architect" by the Architects National Council, in 2015. He states that architecture has to be contemporary in absolute terms, but in a continuous reformulation of exchange relations to the antiques. Changes may be made to these terms from time to time, and the real task consists in reinterpreting the existing condition. Therefore, it is certainly relevant to reread the state of affairs in order to provide new answers, updating the input data so as to generate a fresh output. To sum up, it appears a combination comprising betrayal, translation and renewal refers to the useful life period of a building, whereby the end of a cycle is coupled with its regenerative process. On the other hand, nowadays, it seems that what one considers "high fidelity" is ambiguously associated with a wide range of retail items, such as smartphones or television screen resolutions (1080p, 2K, UHD, 4K, 8K sizes in pixels, and so on). However, the concept of betrayal is actually a cyclical re-actualisation of things past that transform themselves, yet it continues to be stereotyped, especially in western cultural identity.

All things considered, it seems reasonable to assume that, if one wants to solve the most pressing environmental issues, working on consumption is a necessary requisite. For instance, Sweden has recently proposed the legislation that would cut regular tax on repairs of bikes, clothes and shoes from 25% to 12%, but, at the same time, the Scandinavian country is IKEA's homeland, an internationally known store group famous for cheap furniture and home accessories. Only 20 years ago, in a 1996 business advert titled "Chuck Out Your Chintz", IKEA showed British customers rioting against their lifestyles, dumping their chintzy furnishings into a huge skip. Given what I have discussed so far, the unchanged risk appears to be present: ontological propaganda in a disposable world. Fabrizio Aimar

Praz-de-Fort 谷仓改造项目

Savioz Fabrizzi Architectes

该建筑的前身是建造于19世纪后半叶的谷仓，后遭到废弃。它最早位于瑞士Val d'Entremont的Praz-de-Fort村入口处。作为改造过程的一部分，谷仓被拆迁到距原址几公里的Saleinaz山谷中。

在保留了旧谷仓外围护结构的同时，建筑师对其内部进行了彻底的翻新，建造了全新的独立结构及内部表层。内部结构与外部结构分离开来，通过能彼此相互交流的开放式半楼层的设计，形成了丰富的空间组合。因此，可以将整个居住部分看作是一整个开放的连续空间，仅通过高度上的差异来划分不同的功能。建筑师将最为私密的空间——卧室和办公室——设计在谷仓的上部，也就是客厅和厨房的上方。

原有谷仓唯一被保留下来的洞口是通往不同阳台的门，并在门洞内安装上了玻璃。建筑师将立面上新洞口的数量控制在最低限度以内，在提升室内空间舒适性的同时，也避免影响谷仓原有的特性。根据位置的不同，新洞口或与周边的自然环境、或与建筑原有的外围护结构建立了相互协调的关系。

建筑师适当增加了原建筑外部用来晾晒稻麦的走廊上的栏杆，使外墙更加均质并且更接近谷仓最初的外观，同时为室内空间的使用者提供了一种亲密的氛围。

建筑的地基以露石混凝土重新打造，让人想起谷仓最初的石材地基。新的地基通过沿外墙内缘设置的玻璃带与谷仓连接，从而削弱了地基的存在感。玻璃带的设计使得自然光线能够到达地基内部的地下空间。

新谷仓的室内墙面和地面均由橡木板构成，而地基内的空间则采用石材铺面，与之形成了鲜明的对比，这种材料也让人联想起室外的石材。

Conversion, Praz-de-Fort

Built in the second half of the 19th century, the barn was originally situated at the entrance to the village of Praz-de-Fort in the Val d'Entremont, and had been abandoned. As part of the conversion process, it was first dismantled and then rebuilt a few kilometers further into the valley, at Saleinaz.
A new independent structure and an internal skin were constructed inside the envelope formed by the old barn. This separation from the external structure enabled a rich spatial mix to be created via open half-storeys communicating with one another. The dwelling was therefore treated as a single open and continuous space, organised via the differences in level. The bedroom and office, which are the most private spaces, were created in the upper part of the barn, above the living areas and the kitchen.

项目名称：Conversion, Praz-de-Fort
地点：Praz-de-Fort, Valais, Switzerland
建筑师：Savioz Fabrizzi Architectes
合作方：Jean-Pascal Moret, Fabian Wieland
土木工程师：Alpatec Sa, Martigny
供暖工程师：Tecsa Sa, Conthey
供暖：bedrooms, wc, bathrooms, kitchen, living room, terrace, larder, woodshed, plant rooms
用地面积：1,086m² / 建筑面积：101m² / 总建筑面积：232m² / 体积：775m³
设计时间：2014 / 施工时间：2015—2016
摄影师：©Thomas Jantscher (courtesy of the architect)

N

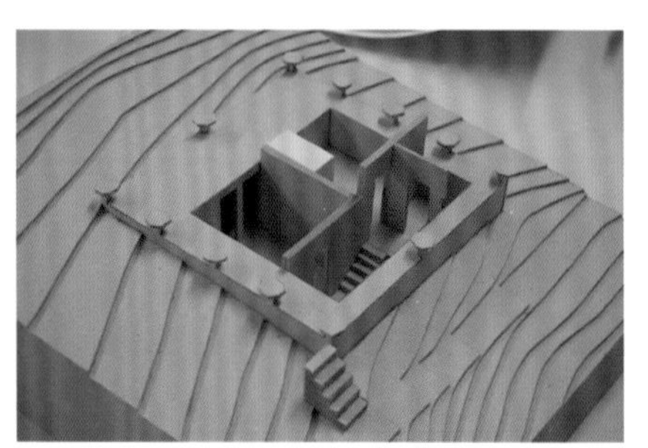

拆迁之前的原先谷仓
original barn before dismantling

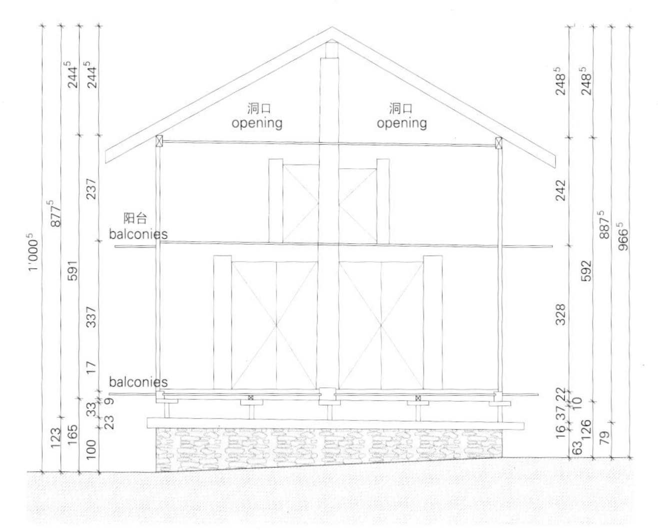

南立面 south elevation

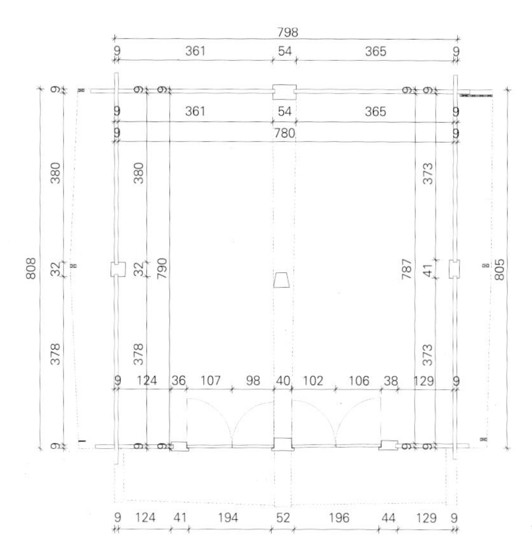

一层 first floor

改造之后
after conversion

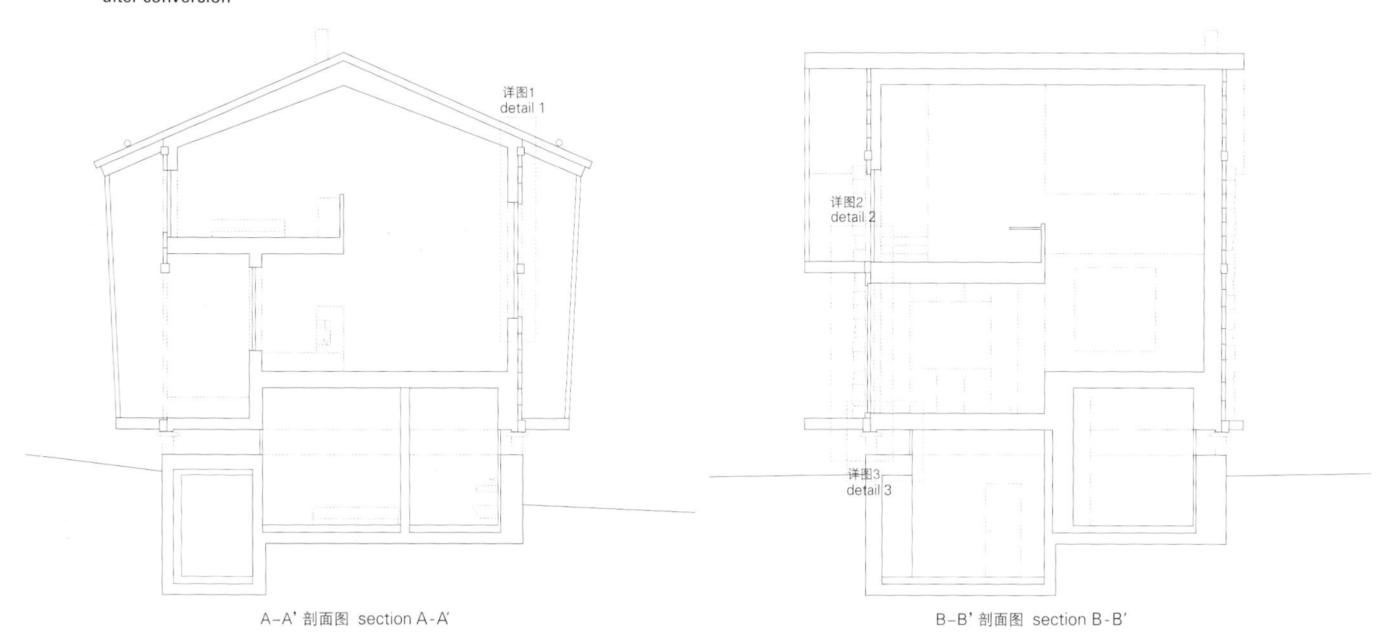

A–A' 剖面图 section A-A'

B–B' 剖面图 section B-B'

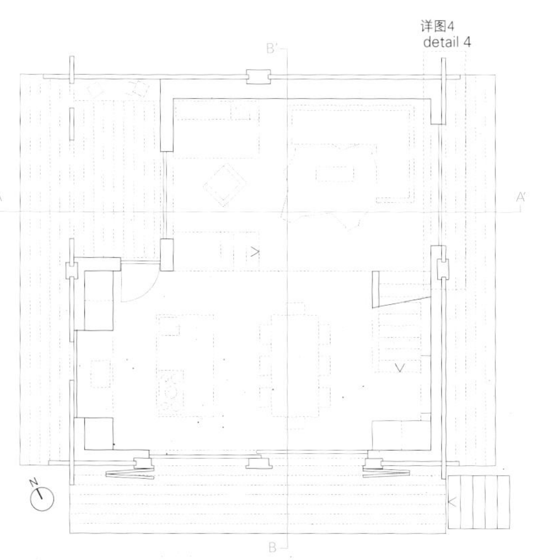

一层 first floor

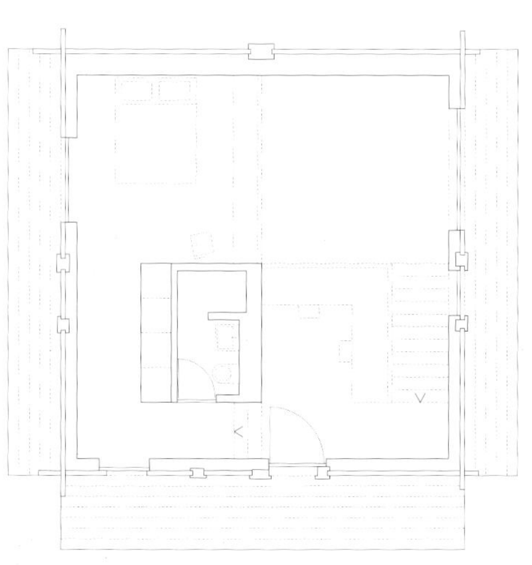

二层 second floor

As the only existent openings were doors that gave access to the different balconies, these were retained and glazed. The number of additional new openings was minimised, to avoid detracting from the character of the barn while making it more comfortable. The new openings were positioned in such a way as to create a relationship either with the natural surroundings outside or with the original envelope, depending on the position.

The external bands formed by the existent ruchines (wheat-drying galleries) have been made slightly more dense with a view to making the outer walls more homogeneous and keeping the barn closer to its original appearance, as well as offering an intimacy for the users of the internal spaces.

The base, which has been reconstructed in exposed concrete, recalls the stone used for the original base. This new base is connected to the barn by a band of glazing which is set back from the external wall so as to make it inconspicuous. This band of glazing enables natural light to reach the areas situated in the base.

The new internal skin of the barn consists of oak panels, with the spaces in the base providing a contrast through the use of materials that recall the external stone.

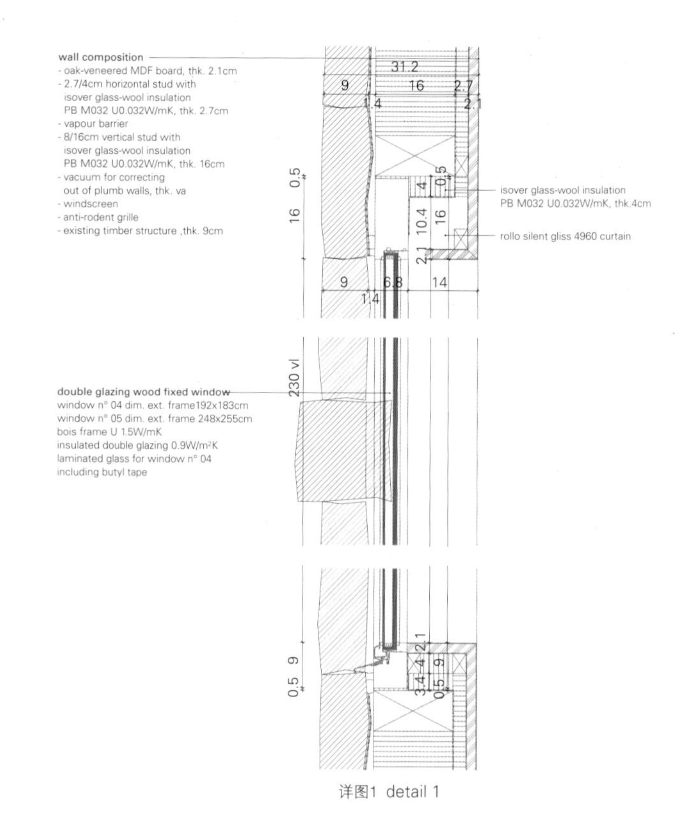

详图1 detail 1

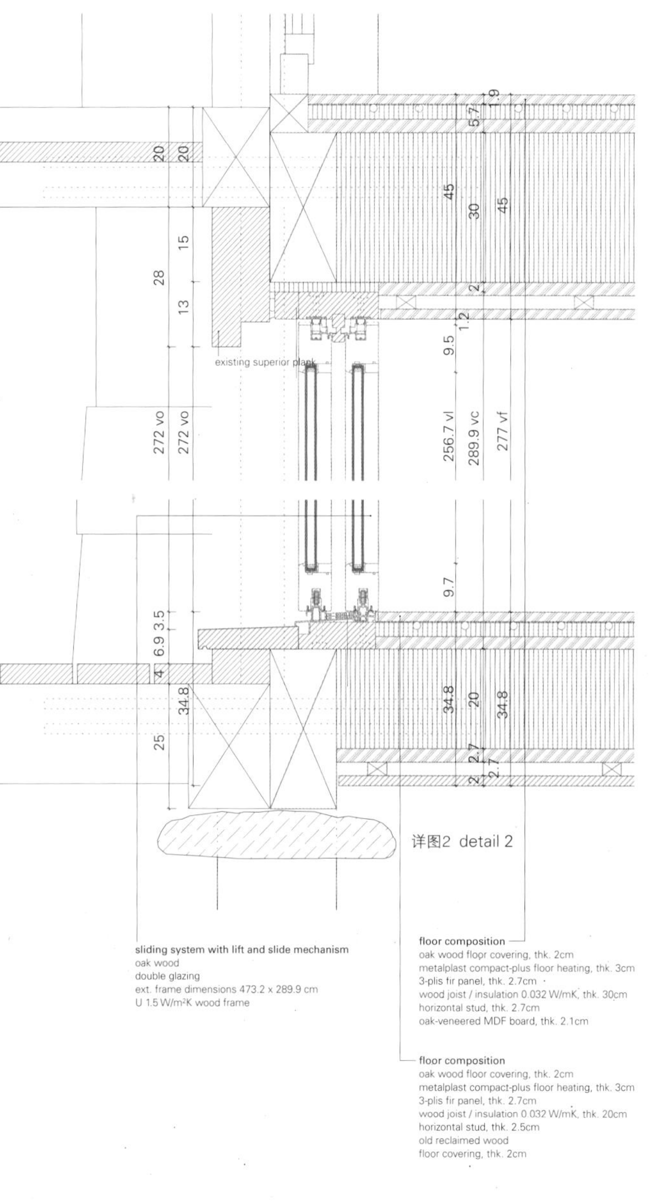

详图2 detail 2

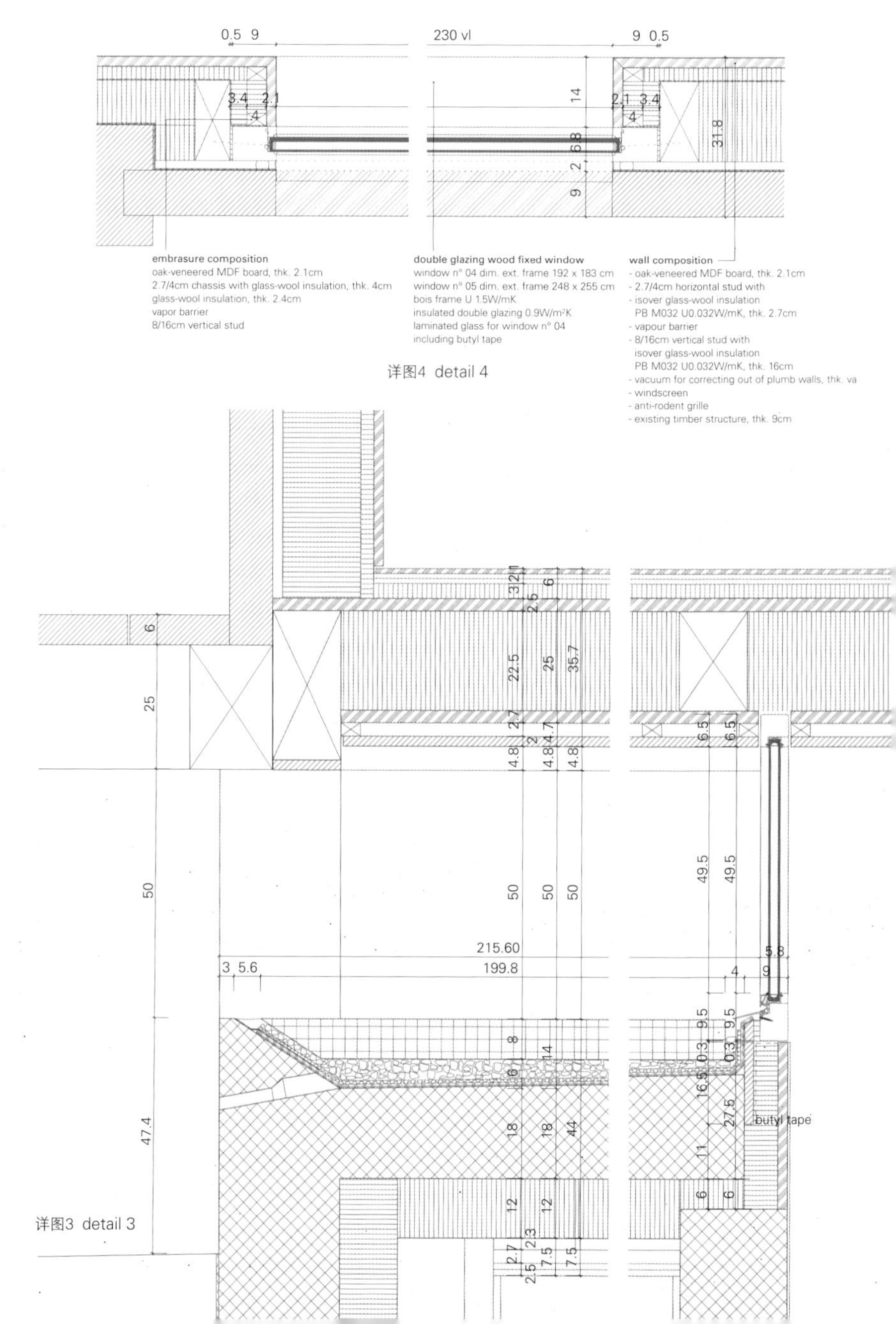

详图4 detail 4

详图3 detail 3

兰科湖畔的谷仓住宅

Estudio Valdés Arquitectos

该谷仓住宅项目是一项回收再利用来自三栋不同建筑的废弃材料的施工提案。本项目是对位于Aculeo泻湖的一处在2010年2月地震中遭到损坏的旧谷仓进行翻新，过程中使用的材料分别来自于一处位于圣地亚哥的烧烤屋、一个位于科尔查瓜山谷的酒店露台和一处位于兰科湖的凉亭。

该项目的主要目标是重新利用旧谷仓遗留下的大部分材料。由白杨木制造的旧谷仓的门和屋顶结构变成了桁架和包板，而从当地采购来的木料则被用于房屋的支柱和外部覆层。建筑的结构由回收再利用的桁架组成，其周界每隔三米设置的支柱为其提供支承力。如此一来便形成了一个简洁的建筑体量，带有一个单独的室内大厅，无需添加任何结构。

这座住宅（避暑别墅）的设计方案以公用起居空间为中心。露台、起居室和餐厅位于主厅，而其他功能区，如卧室、浴室和厨房以及设备区，都聚集于谷仓的一角，形成了一个L形的布局。

根据性能，每一个空间采用的材料都各不相同。公用空间完全透明，将一排排由热板玻璃包裹的桁架和立柱的肋材露在外面。

另一方面，卧室和设备区采用不透明设计，小屋的屋顶为其提供庇护，并且整体包覆了回收再利用的白杨木。

最初，谷仓位于林中一处天然高地之上，如今，它变成了轻轻安放于一排立柱之上的巨大屋顶。鉴于立面的透明设计，该住宅随着日升月落始终在发生变化。白天，住宅另一侧的树木使得室内空间融入了外部的树林；而在夜晚，这座谷仓又变成了一座灯塔，在树干和树枝之间散发着光芒。

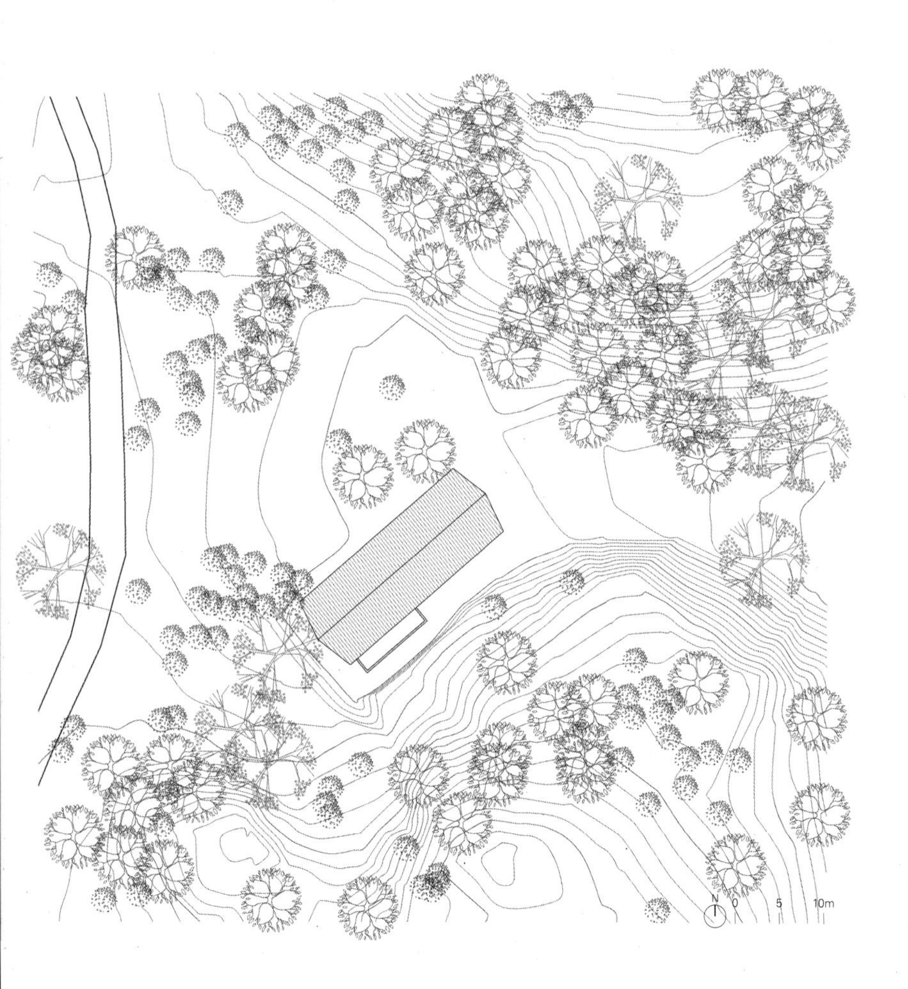

Barn House at Lake Ranco

The Barn House is a construction proposal recycling the materials salvaged from three different buildings. Renovating an old barn located in the Aculeo lagoon that was damaged by the earthquake of February 2010, the materials used come from a barbecue house in Santiago, a hotel terrace in the Colchagua Valley, and a summerhouse at Lake Ranco.
The main objective was to reuse most of what remains of the old barn. The doors and the roof structure in poplar wood were turned into trusses and sheathing, while wood from local wood sources were used in pillars and exterior coatings. The structure consists of recycled trusses supported by columns standing every three meters around the perimeter of the building. This shapes a simple and clean volume that has a single interior hall exempt of any structural obtrusion.
The program of the summer house centers on the common living spaces. The terrace, living room and dining room are situated in the main hall, gathering the rest, such as bedrooms, bathrooms, the kitchen and service area, all together at a corner of the barn forming an " L" shape arrangement.
The materiality of each space differs by its properties. The common area is completely transparent, exposing the sequential ribs of trusses and pillars enclosed with thermo panel glass. On the other hand, the bedrooms and service area are opaque, sheltered under the roof of the shed and entirely cladded with the recycled poplar wood.
The barn strategically situated on a natural plateau in the middle of the forest, now becomes a great cover resting gently on a row of pillars. Through its transparent façade, the house changes by day and night. In daylight, the trees on the other side sets the interior space amidst the woods outside. At night the barn becomes a light house glowing among the trunks and branches.

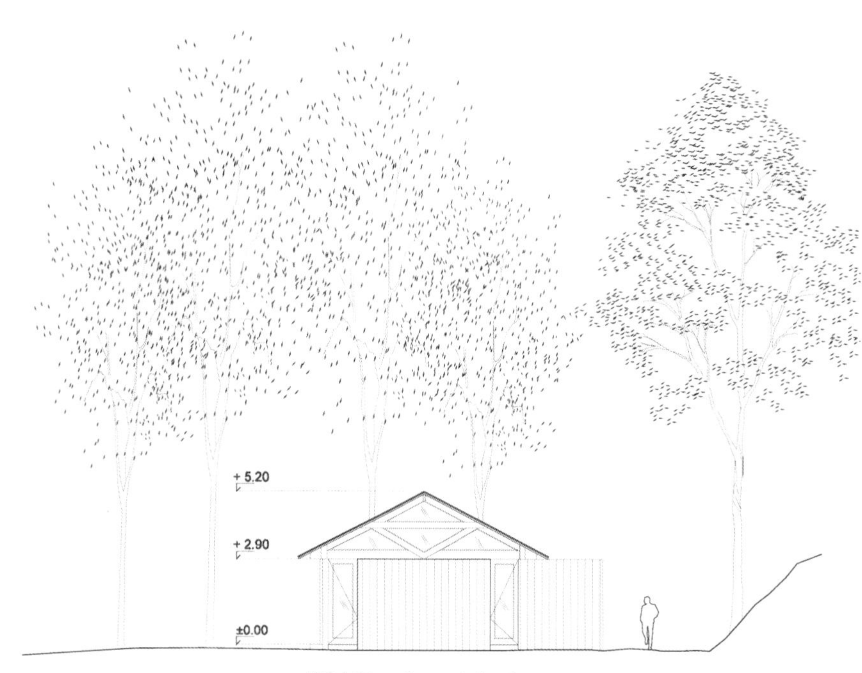

西南立面 south-west elevation

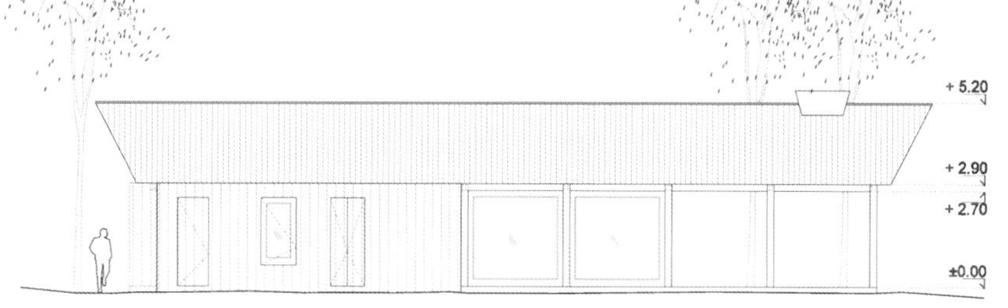

东南立面 south-east elevation

东北立面 north-east elevation

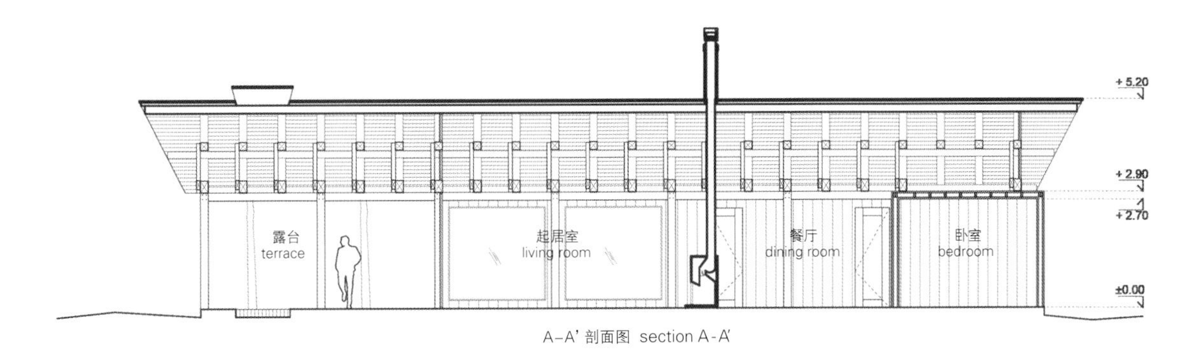

A–A' 剖面图 section A-A'

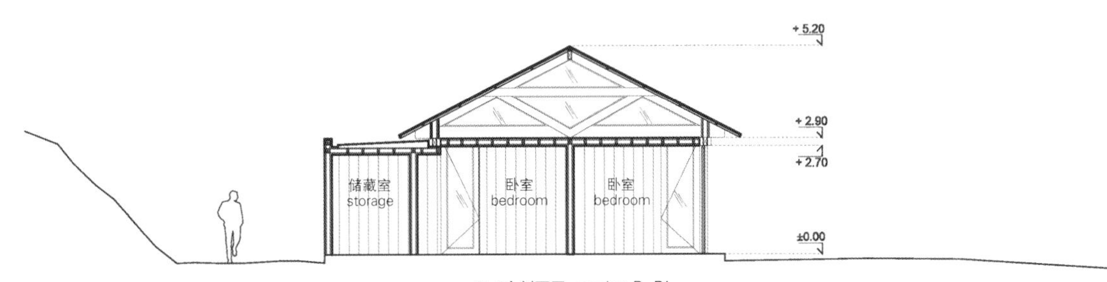

B–B' 剖面图 section B-B'

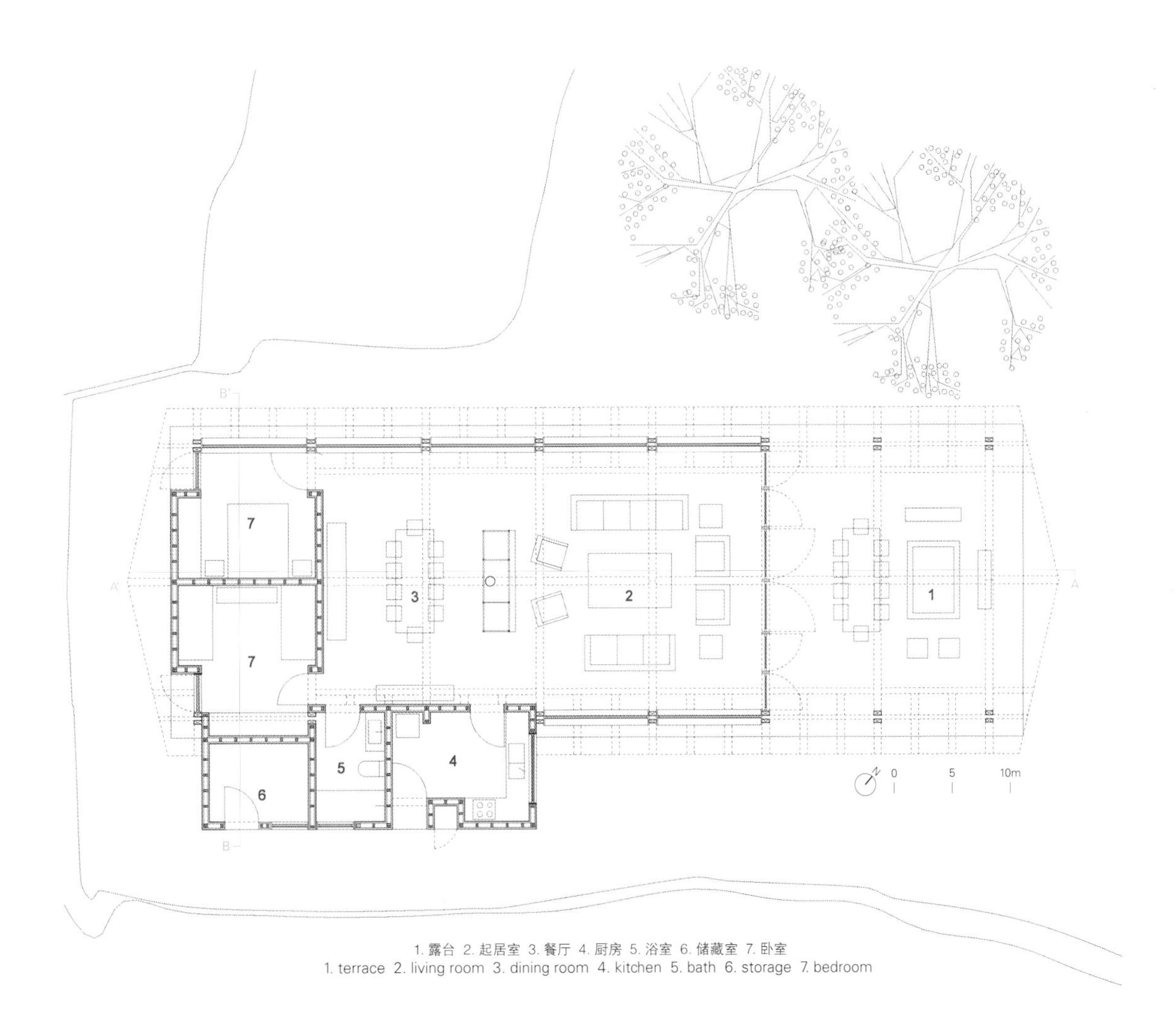

1. 露台 2. 起居室 3. 餐厅 4. 厨房 5. 浴室 6. 储藏室 7. 卧室
1. terrace 2. living room 3. dining room 4. kitchen 5. bath 6. storage 7. bedroom

项目名称：Barn House at Lake Ranco / 地点：Lago Ranco, Región de los Ríos, Chile /建筑师：Estudio Valdés Arquitectos
主管建筑师：Leonardo Valdés Cruz, Carlos Ignacio Cruz Elton, Alberto Cruz Elton / 合作方：Isabel de la Fuente
结构工程师：Alberto Ramirez / 建筑面积：inside; 138m² + terrace; 54m² / 材料：windows; glass / coating; recycled wood, poplar wood / structure; poplar wood / 造价：$600 US/m² / 施工时间：2014—2015
摄影师：©Felipe Díaz Contardo (courtesy of the architect)

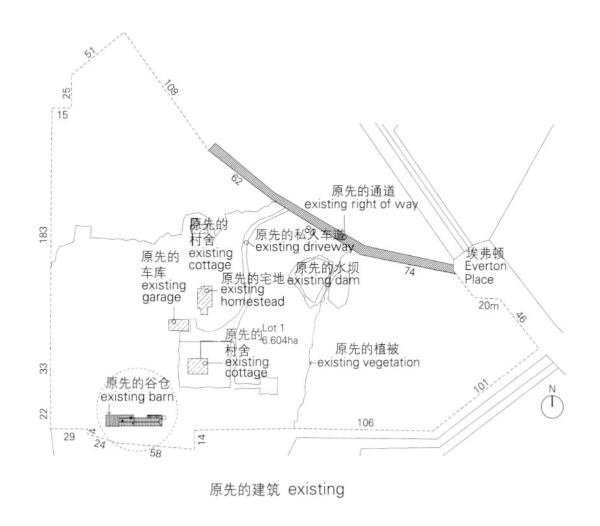

原先的建筑 existing

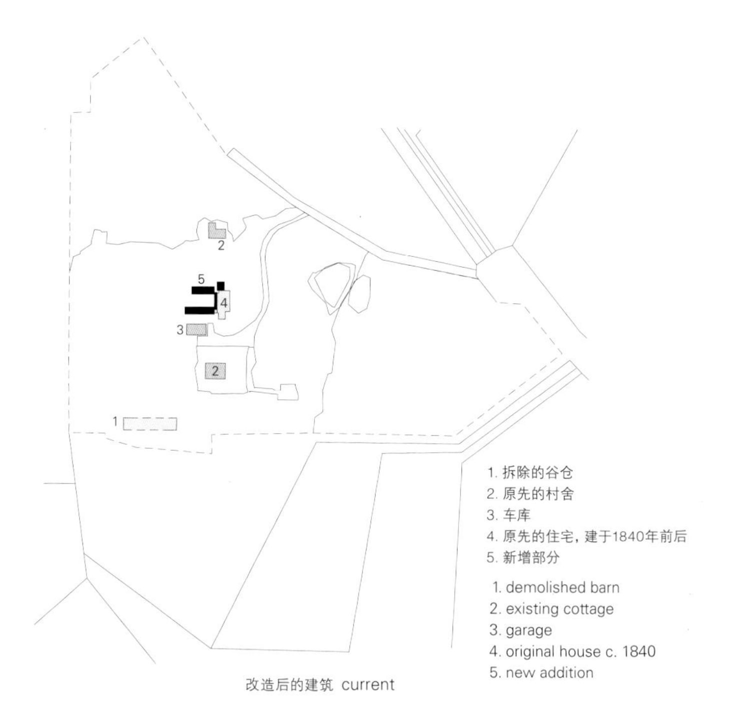

改造后的建筑 current

谷仓改造

Circa Morris-Nunn Architects

随着时间的推移，所有的房屋都会经历演变的过程。距今近200年的阿克顿住宅，从一所早期殖民时期庄园主的宏伟（而又质朴的）住所（带有仆人房间和关押罪犯的牢房，同时作为附近以家庭为单位的农田劳作者的一个基地）变成了一处单一的家庭住宅，新住宅中的生活模式是当初建造房屋时从来没有设想过的。

扩建建筑所要求的设计方法是建造全新的"附属建筑"，在风格上与宅地后方附近的开放庭院周围的原有建筑相符。扩建材料使用的是石料和木料，来自于已被封闭的最初建于1820年的谷仓，这个谷仓在一场暴风雨中遭受了不可挽回的损失，不但一侧塌陷，屋顶也坍塌了。其木料不是采用扁斧劈开就是用锯切开，包括源自殖民时期的最长的木料（18m）。在施工过程中，建筑师煞费苦心地将这些材料拆卸并重新组装上。

最终建成的建筑成了一种明智的建筑风格之间的对话，这个新项目带有明显的现代特色，但同时又配备了真正来自于殖民时期的建筑构件；扩建建筑的设计效果创造了一种令人感到震撼的视觉冲击力。作为对殖民时期宅地演变的一种阐述，它的意义远比一栋简单的新扩建建筑要丰富得多。

阿克顿住宅堪称"活的"文化遗产，通过响应变化的需求而发生演变。它可以说是举世无双的。

Re [Barn]

All houses evolve over time. Acton, which is almost 200 years old, has changed from a grand (but modest) early colonial manor house with servants/convicts, which also served as a base for the families working on the surrounding rural prop-

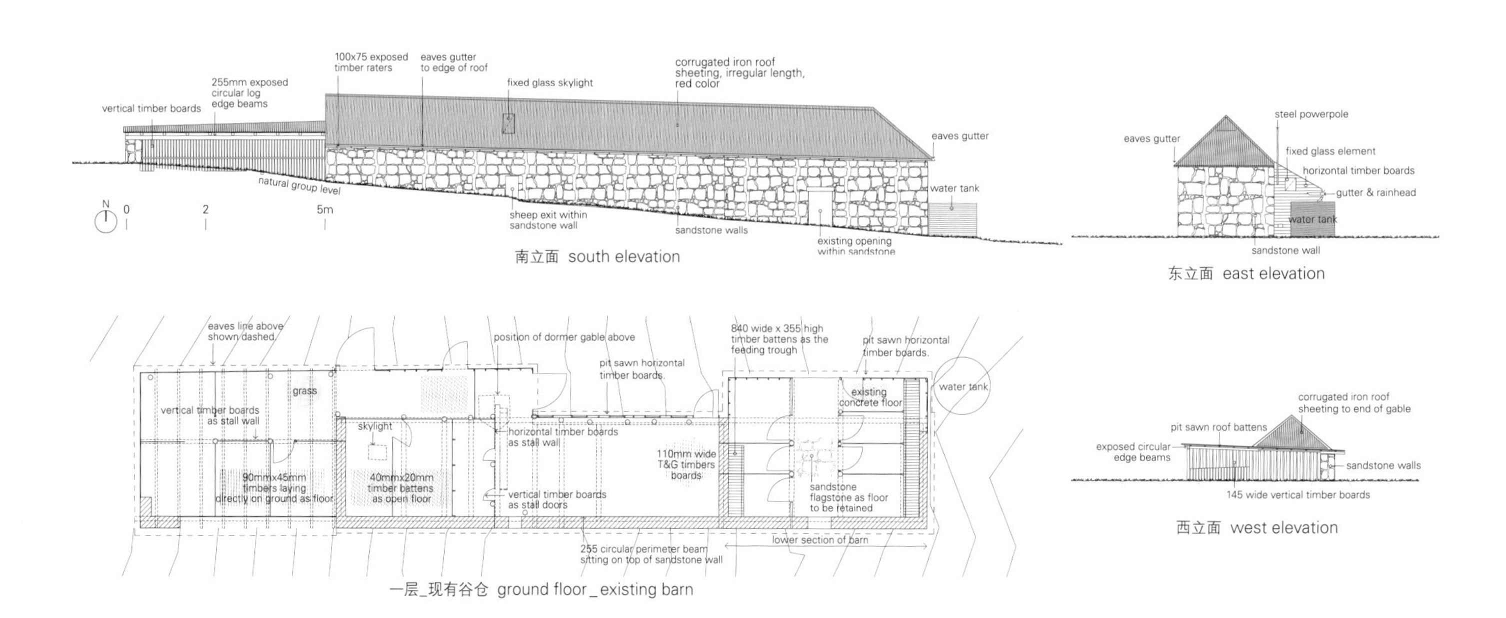

南立面 south elevation

东立面 east elevation

一层_现有谷仓 ground floor _ existing barn

西立面 west elevation

北立面 north elevation

西立面 west elevation

南立面 south elevation

东立面 east elevation

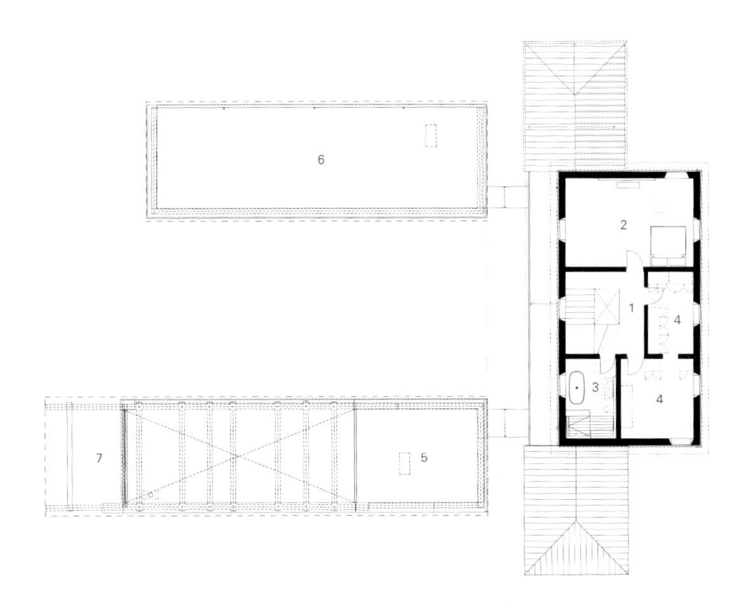

1. 大厅 2. 现有的主卧 3. 现有的浴室 4. 步入式衣橱 5. 储藏室 6. 屋顶空间 7. 带顶天井
1. hall 2. existing master bedroom 3. existing bathroom
4. walk-in robe 5. storage 6. roofspace 7. covered patio
二层 first floor

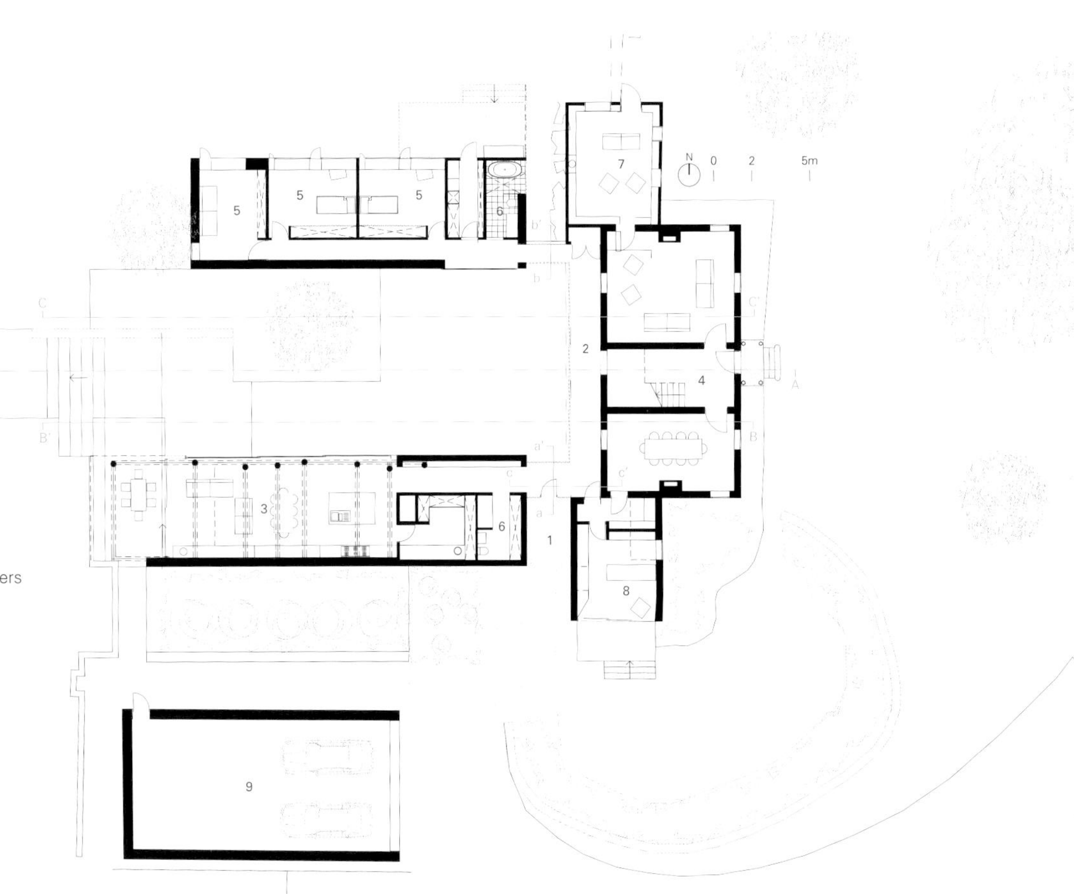

建筑师：Robert Morris-Nunn, Ganche Chua
项目团队：Hannah Gora
结构/设备/机械/电气工程师：JMG engineers and planners
能源顾问：Steve Watson
景观设计师：Play Street
遗产顾问：Peter Spratt
总承包商：Bennett Constructions
摄影师：©Matt Sansom (courtesy of the architect)

1. 入口 2. 玻璃连廊 3. 起居室侧室 4. 原先住宅，建于1840年前后 5. 卧室 6. 浴室 7. 图书室 8. 办公室 9. 现有的车库
1. entry 2. glass linkway 3. living wing 4. original house c. 1840
5. bedroom 6. bathroom 7. library 8. office 9. existing garage
一层 ground floor

erty, to a single family home with a mode of living which could never have been imagined when the house was built.

The design approach to the extensions required was to create new "outbuildings" replicating in spirit those traditionally located around an open courtyard close to the rear of the homestead. The materials used were the stone and timber of the original condemned 1820 barn that had suffered irreparable damage in a violent storm, with one side collapsing and the roof caving in. Its timber was either all adzed or sawn, and included the longest piece of timber(18m) from colonial times. The materials were painstakingly dismantled and rebuilt.

The result is architecture as an informed dialogue, with the new work being obviously contemporary yet with truly ancient colonial components; the result creating something visually compelling. It is far, far richer as a story of the evolution of a colonial homestead than a simple new extension that could ever have been.

Acton is "living" cultural heritage, evolving sympathetically in response to changing needs. It's simply unique.

项目名称：Re [Barn] / 地点：Acton, Tasmania
用途：private residence
用地面积：66,040m² / 总建筑面积：535m²
现有建筑面积：160m² / 新增部分面积：250m²
建筑覆盖率：0.43%
建筑规模：existing building; 2 storeys with an underground cellar, new addition; single storey on the ground
结构：recycled hand adzed timber from demolished barn on site
外部饰面：custom orb roof sheeting, recycled sandstone from demolished barn on site, horizontal timber boards cedar bevel edge cladding-stain finish, double glazed fixed glass
内部饰面：floor; burnished concrete, walls; plasterboard, recycled timber
设计时间：2013 / 施工时间：2014 / 竣工时间：2014

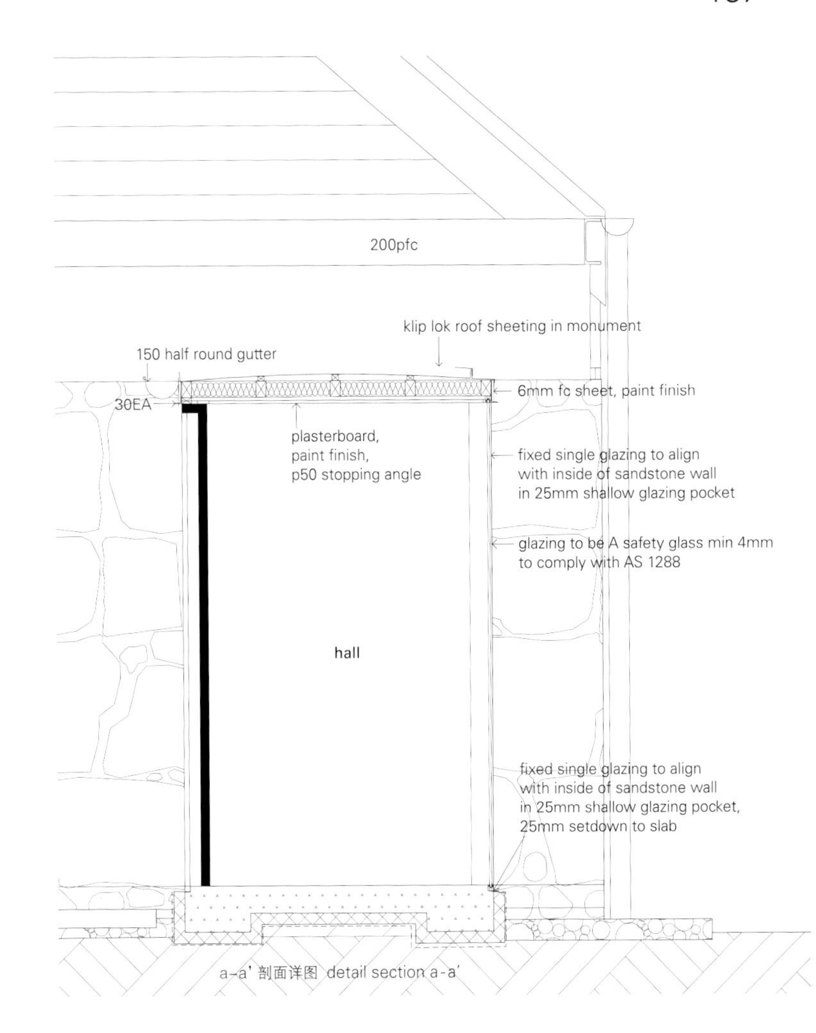

a-a' 剖面详图 detail section a-a'

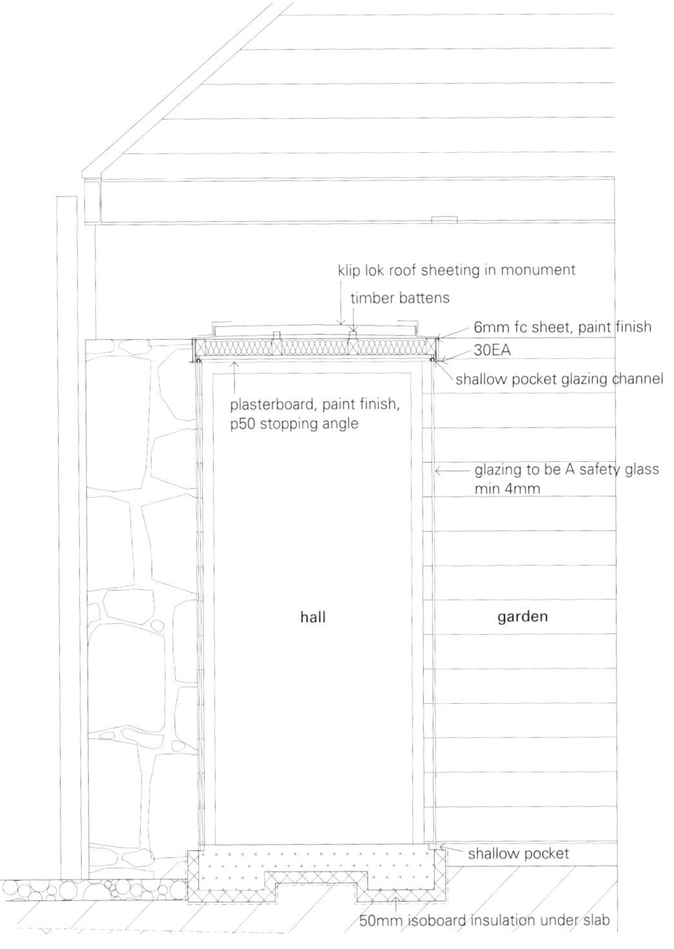

b-b' 剖面详图 detail section b-b'

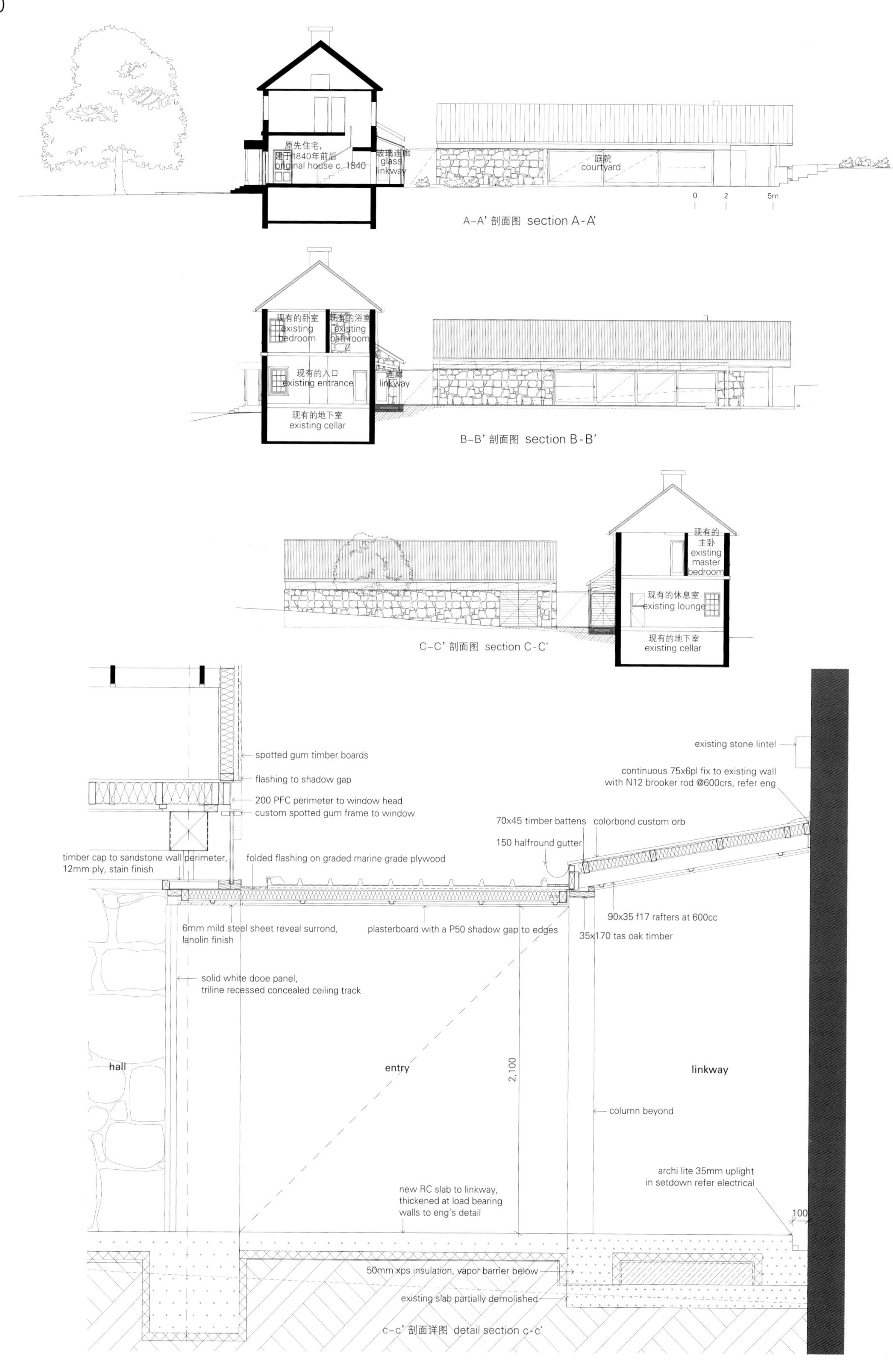

原先住宅，建于1840年前后
original house c. 1840
玻璃连廊
glass linkway
庭院
courtyard
0
2
5m
A-A' 剖面图 section A-A'
现有的卧室
existing bedroom
现有的浴室
existing bathroom
现有的入口
existing entrance
连廊
linkway
现有的地下室
existing cellar
B-B' 剖面图 section B-B'
现有的主卧
existing master bedroom
现有的休息室
existing lounge
现有的地下室
existing cellar
C-C' 剖面图 section C-C'
spotted gum timber boards
flashing to shadow gap
200 PFC perimeter to window head
custom spotted gum frame to window
existing stone lintel
continuous 75x6pl fix to existing wall with N12 brooker rod @600crs, refer eng
70x45 timber battens
colorbond custom orb
150 halfround gutter
timber cap to sandstone wall perimeter, 12mm ply, stain finish
folded flashing on graded marine grade plywood
6mm mild steel sheet reveal surround, lanolin finish
plasterboard with a P50 shadow gap to edges
90x35 f17 rafters at 600cc
35x170 tas oak timber
solid white dooe panel, triline recessed concealed ceiling track
hall
entry
2,100
linkway
column beyond
archi lite 35mm uplight in setdown refer electrical
100
new RC slab to linkway, thickened at load bearing walls to eng's detail
50mm xps insulation, vapor barrier below
existing slab partially demolished
c-c' 剖面详图 detail section c-c'

EXPLORE AUSTRALIA
One People
THE TRAVEL BOOK
the new international atlas
ITALIAN FEAST
ITALY

Kyodo 住宅

SANDWICH + Tam Low-energy

这是一栋以高效利用自然能源为设计主旨的住宅。以一间（当太阳在正南时）朝阳的客厅为中心，住宅的设计使用了一系列叠加的箱型结构，以组织室内空间。冬季里，在阳光的照射下，温暖的空气在整个空间中延伸，而在夏季里，凉爽的微风从窗户吹入室内。这栋住宅有一间书房，起到了通风塔的作用，并且住在这里的人可以透过书房的窗户感受周围树木的绿茵。在这栋住宅中，根本不需要空调。

外墙的独特构成的灵感源自于雕塑家名和晃平创作的探究重力视觉体现的《方向》系列画作。最初从生长于日本的不同树木上砍伐下来并遗忘在木料场的旧木料被重新利用，排布起来，所以木料的宽度和颜色交替排列。这样设计的目的在于使建筑带给人一种温馨、淳朴而又带有现代风格的印象。

项目的平面布局是按照单一的家庭住宅或合租房的形式来构思的。设计师在起居空间的中庭设计了一些大书架，作为业主生活方式的中心装饰品，并且中庭有一处凹入地面之下的代用空间，可用于开展各种活动。鉴于这样的设计，此住宅中的人能够与当地社区产生联系，同时分享了结合艺术、永久培养及相似方法的城市生活新方式。

Kyodo House

A house design to make effective use of natural energy. Centered on a living room oriented to face the sun when it passes due south, the program of this house uses a number of overlapping boxes to organize the interior space. In winter, air warmed by the sunlight extends throughout the space, while in summer, a cool breeze from the windows passes through. The house has a study room that functions as a ventilation tower, and from its windows you can feel the greenery of sur-

rounding trees. Air conditioners are not needed in this house. The distinctive composition of the external wall is based on the concept of a sculptor Kohei Nawa's "Direction" series of paintings, which explores a visualization of gravity. The old timber that was originally cut from different trees growing in Japan and left out in the timber yard is reused, arranged so that width and colors alternate. The aim was to make an impression that is warm and rustic while still being contemporary.

The layout was conceived as a single family home or a share house. The central atrium in the living space has large bookshelves that are the centerpiece of owner's living style, and there is an alternative space sunken into the ground for events. This house enables people who live there to connect with the local community, while sharing new ways of city lifestyle that incorporates art, permaculture, and similar approaches.

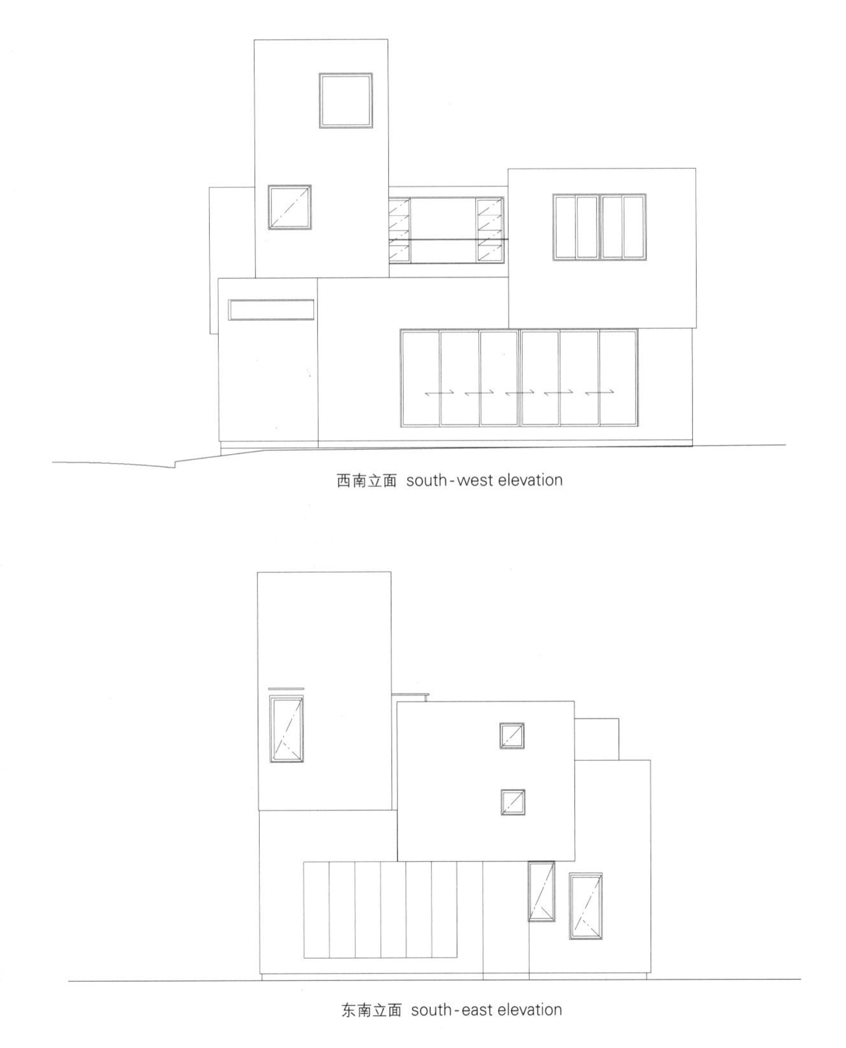

西南立面 south-west elevation

东南立面 south-east elevation

NO
SMOKING
KONDO

空气循环示意图
air circulation diagram

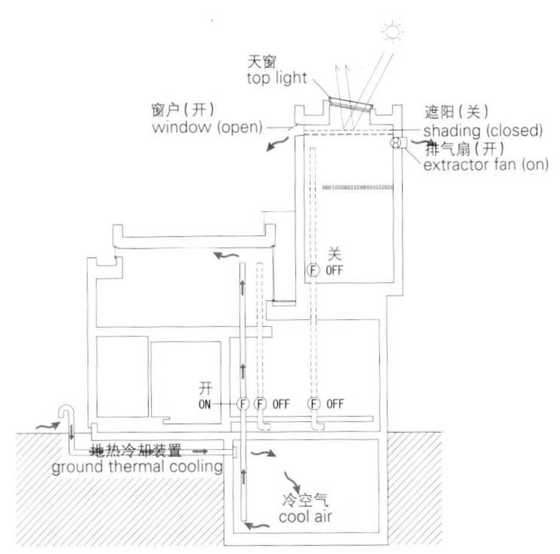

夏季 summer

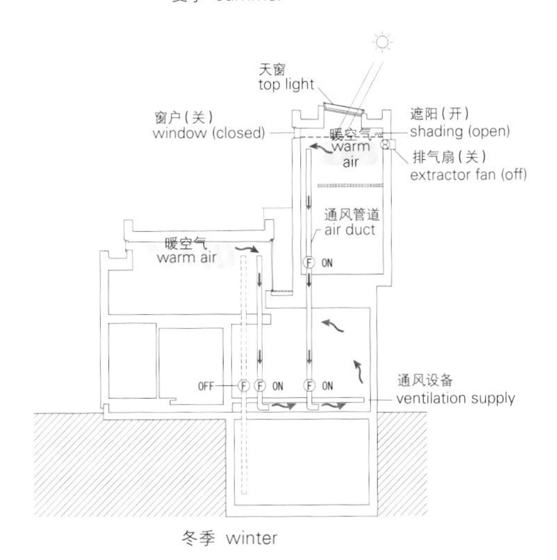

冬季 winter

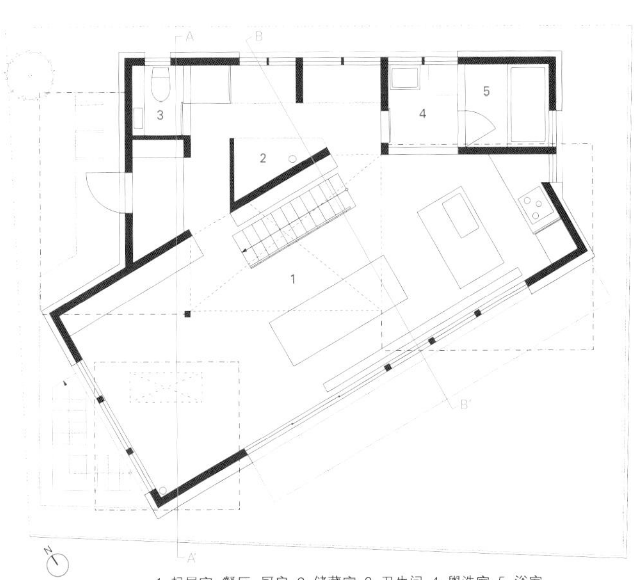

1. 起居室、餐厅、厨房 2. 储藏室 3. 卫生间 4. 盥洗室 5. 浴室
1. living, dining, kitchen 2. storage 3. toilet 4. wash room 5. bathroom
一层 ground floor

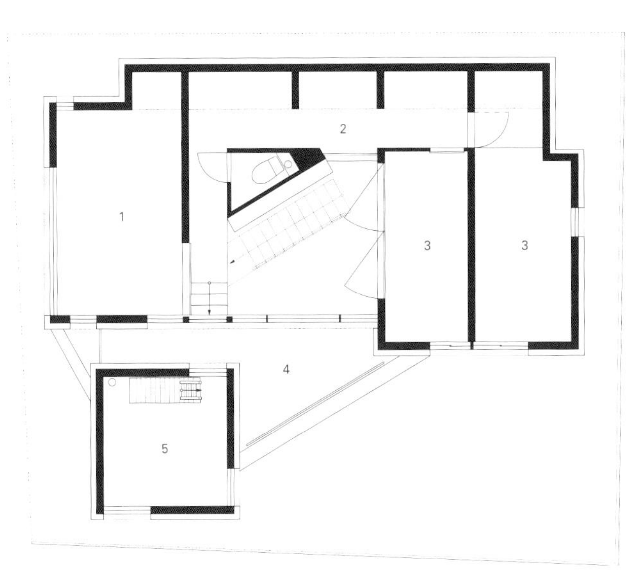

1. 卧室 2. 储藏室 3. 房间 4. 露台 5. 工作室
1. bedroom 2. storage 3. room 4. terrace 5. work room
二层 first floor

项目名称：Kyodo House
地点：Setagaya-ku, Tokyo, Japan
建筑师：SANDWICH + team Low energy
结构工程师：Hiroshi Ono
设备工程师：ymo
照明设计：Plus y
承包商：Ohara Construction
用途：dwelling
用地面积：129m²
总建筑面积：156m²
竣工时间：2015.6
摄影师：courtesy of the architect

A-A' 剖面图 section A-A'

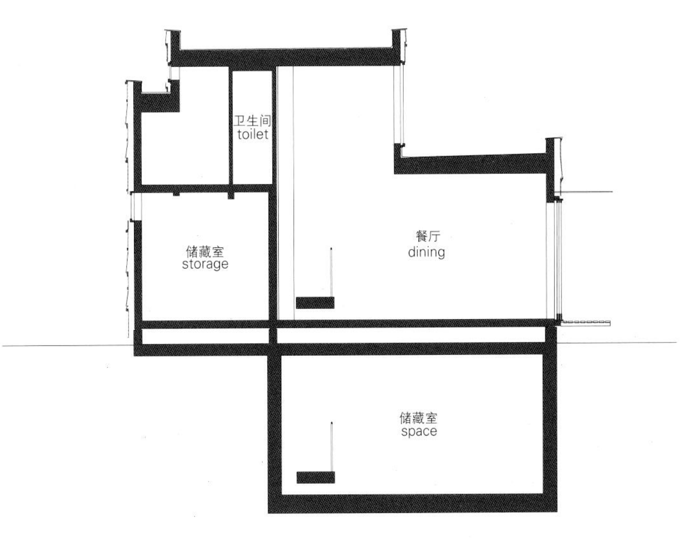

B-B' 剖面图 section B-B'

P120

P144

P120 Steyn Studio

London based Steyn Studio was founded by South African born Coetzee Steyn[p.202-top] in 2011, after leaving behind a career in the one of the biggest commercial practices in London, in order to seek a more humanely direct connection in the architecture that he would like to produce. He took an unusual route to becoming an architect as he started his education at the Cape Peninsula University of Technology, with the aim to first truly understand the technicality of architecture, prior to the design and theory biased disciplines he later followed at the University of Cape Town in South Africa and later the Architectural Association in London, UK where he qualified as a professional architect.

P20 BIG

Was founded in 2005 by Bjarke Ingels and now based in Copenhagen and New York. The group of architects, designers, builders, and thinkers operats within the fields of architecture, urbanism, interior design, landscape design, product design, research and development. The office is currently involved in a large number of projects throughout Europe, North America, Asia and the Middle East. Their architecture emerges out of a careful analysis of how contemporary life constantly evolves and changes.

P110 Archi-Union Architects

Is a Shanghai-based architectural design firm, founded in 2003. Has a Grade A design qualification certificate issued by The Ministry of Housing and Urban-Rural Development. Its business scope includes comprehensive design services such as architectural design, urban planning, landscape design and interior design. Adheres to the combination of traditional Chinese culture and digital construction technology, and practices the symbiotic relationship of nature, urban and architecture.

Founder, Philip F. Yuan[p.203-bottom] is a Ph.D Advisor in the College of Architecture and Urban Planning (CAUP), Tongji University. Is also a Council Member of Architects Sector and Digital Fabrication Sector of the Architectural Society of China.

P144 Localarchitecture

Was founded by Manuel Bieler[p.202-bottom, left], Antoine Robert-Grandpierre[p.202-bottom, center] and Laurent Saurer[p.202-bottom, right] in Lausanne, Switzerland in 2002. Manuel Bieler received a M.Arch in 1996 and attended a Postgraduate course of Construction Economy and Managment at the EPFL. Ántoine Robert-Grandpierre also received a M.Arch in 1996 at the EPFL. Is President of the Maison de l'Architecture, Geneva since 2013. Laurent Saurer has worked at the Geninasca & Delefortrie Architects, MPH Architects and Workshop 15X10. Is a Member of the Commission of the Federation of Swiss Architects since in 2013. They are Member of the Federation of Swiss Architects (FSA) since 2010.

P98 Kubota Architect Atelier

Was established by Katsufumi Kubota[p.203-top] in 1988. He was born in Yamaguchi, Japan in 1957 and graduated from Nihon University in 1981. Was a part-time lecturer at Hiroshima Institute of Technology and Yamaguchi University. Received numerous architectural honors including Barbara Cappochin International Prize 2005, Good Design Award 2005, JCD Design Award 2006, Chicago International Architecture Award 2006 and so on.

P68 archipelagos - Manolo Ufer

Is an architecture studio specialized in experimental housing and participates in design-and-build projects, landscape and infrastructural development, strategic planning and urbanism. Currently operates in the United States, Mexico and Spain.
Manolo F. Ufer was born in Madrid, Spain in 1976 and studied architecture at the University of Edinburgh. Received an AA Honours Diploma at the AA School in London in 2002 and worked in Europe and America. Established his own architectural practice, archipelagos in 2007.

P56 Anako'architecture

Anako is the name of an ancient Indian shaman. The habitants of Anako lived in perfect harmony with nature and their homes were seamlessly integrated into Native American forests. The architecture of Anako has the same mission: to create a habitat where the place is respected and maintained. On the other hand, today, the space of life and work adds a new dimension. The buildings seem to be an extension of ourselves and we must look towards comfort and that's where Anako'architecture pay attention.

P130 Office KGDVS

Was founded in 2002 by Kersten Geers[left] and David Van Severen[right]. Both of them graduated in Architecture and Urbanism at the University of Ghent, Belgium and at the ETSA Madrid, Spain. Kersten Geers worked with Maxwan Architects and Urbanists and with Neutelings Riedijk Architects in Rotterdam. Was professor at the University of Ghent, and visiting professor at Columbia University, NYC, and the Academy of Architecture in Mendrisio, and is currently teaching at the EPFL, Lausanne. Is a founding member of the architecture magazine *San Rocco*. David Van Severen worked with Stephane Beel Architects, Xaveer De Geyter Architects and for Atelier Maarten Van Severen. Was a teacher and guest critic at the Academy of Amsterdam, the TU Delft, and the Berlage Institute in Rotterdam, and is currently guest tutor at Architecture School of Versailles.

P192 SANDWICH

Was established by Kohei Nawa, Yoshitaka Lee and Yuichi Kodai in 2013. Kohei Nawa was born in Osaka, 1975. Received PhD of Fine Arts from the Kyoto City University of Arts in 2003 and established SANDWICH in Kyoto in 2009. Currently is a Professor at Kyoto University of Art and Design. Yoshitaka Lee was born in Osaka 1967. Graduated from the Osaka University of Arts Junior College in 1988. Established atelier Lee in 2005. Yuichi Kodai was born in Kyoto and graduated from the University of East London in 2001. Completed a Diploma Course at the Architectural Association in 2005.

P192 Team Low-energy

Nihei Wataru was born in Tokyo and graduated from the Hosei University in 1983. Has worked at the Plants Associates Inc. and founded team Low-energy in 2015. Currently lives and works in Tokyo.

Fabrizio Aimar

Graduated cum laude from Politecnico di Torino(Italy). Worked for five years for a civil and infrastructural engineering office in Torino. Has developed different structural projects in collaboration with some international architectural firms such as Jean Nouvel, Renzo Piano, Mario Cucinella and Aymeric Zublena. Since 2010, he has been a contributor for the Italian architectural magazine *Il Giornale dell'Architettura* and also a member of Cultural Committee of the Asti local board of Architects. Since 2014, he is an external writer for the Italian technical websites "architetto.info" and "ingegneri.info".

Nelson Mota

Graduated at the University of Coimbra, where he lectured from 2004 until 2009. Currently is an Assistant Professor at the TU Delft, in the Netherlands, where he concluded in 2014 his PhD with the title "An Archaeology of the Ordinary. Rethinking the Architecture of Dwelling from CIAM to Siza". Was the recipient of the Távora Prize (2006) and authored the book *A Arquitectura do Quotidiano* (2010). Is a member of the editorial board of the academic journal *Footprint*, and a founding partner of Comoco architects.

P184 Circa Morris-Nunn Architects

Director and principal architect, Robert Morris-Nunn graduated from the Sydney University in 1972 and received M.Arch at the Royal Melbourne Institute of Technology (RMIT). Is regarded as one of Tasmania's most adventurous architects. Has practised in Tasmania for over 25 years, and taken a special interest in the social impact of architecture and collaborative design processes. Has designed and built several groundbreaking buildings, which have received great international interest in the field of ecologically sustainable design. Was appointed as an Adjunct Professor at the School of Architecture, University of Tasmania in 2009. Is a Fellow of the Australian Institute of Architects.

P34 Herzog & de Meuron

Was established in Basel, 1978. Has been operated by senior partners; Christine Binswanger, Ascan Mergenthaler and Stefan Marbach, with founding partners Pierre de Meuron and Jacques Herzog. An international team of about 40 Associates and 380 collaborators is working on projects across Europe, the Americas and Asia. The firm's main office is in Basel with additional offices in Hamburg, London, New York City, and Hong Kong. The practice has been awarded numerous prizes including the Pritzker Architecture Prize (USA) in 2001, the RIBA Royal Gold Medal (UK) and the Praemium Imperiale (Japan), both in 2007. In 2014, awarded the Mies Crown Hall Americas Prize (MCHAP).

P172 Estudio Valdés Arquitectos

Alberto Cruz Elton was born in Santiago, Chile in 1975 and studied at Pontifical Catholic University of Valparaíso. Attended various international workshops between 1993 and 1995 and received his architectural degree from Finis Terrae University in 2000. Has worked at the Schmidt y Valdés Arquitectos Asociados after graduation. Carlos Ignacio Cruz Elton was born in Santiago, Chile in 1978. Was granted the Architecture Honor Student Scholarship in 2003 and graduated from Finis Terrae University with the same Honors in following year. Has worked at Rafael Viñoly in New York from 2005 and returned to Chile in 2006 to become a Partner in Estudio Valdés.

Julia van den Hout

Was born in Amsterdam and moved to New York in 2004, where she attended New York University. After graduating with a Bachelors degree in Art History and Urban Design, she started work at Steven Holl Architects, where she is currently the Press and Exhibitions Manager, handling press for both the New York and Beijing office. She is involved in studies at the Design Criticism department at the School of Visual Arts, with a focus on architecture criticism and Dutch Design.

P82 IDMM Architects - HeeSoo Kwak

HeeSoo Kwak studied architecture at the HongIk University and established IDMM Architects in 2003. Has been contributed a column based on the urban issues in the JoongAng Ilbo from 2016. Received 2016 American Architecture Gold Prize, 22nd and 24th WA(World Architecture Community Award), 39th Korean Institute of Architects Award (Best 7 Award) and Grand Prize of the Korean Architecture Award 2016. His major works include 42nd Root House, Tethys, and Moken Pension.

P162 Savioz Fabrizzi Architectes

Founded in 2004 by the two architects, Laurent Savioz[right] and Claude Fabrizzi[left], is trying to respond with the best conditions to the needs of the clients by providing all the architectural services from the project to the achievement. Their work is based on the analysis of a site in its natural or built state in order to identify the essentials elements that could enhance, preserve or qualify a site. In this way, the firm enhances the cultural role of the architecture based on the analysis of a function, respectively a program, its place in the history and the culture of a region.

图书在版编目(CIP)数据

复兴混凝土 / 丹麦BIG建筑事务所等编 ; 蒋丽，周茎译. -- 大连 : 大连理工大学出版社，2018.3
(建筑立场系列丛书)
ISBN 978-7-5685-1385-2

Ⅰ. ①复… Ⅱ. ①丹… ②蒋… ③周… Ⅲ. ①混凝土一研究 Ⅳ. ①TU528

中国版本图书馆CIP数据核字(2018)第042066号

出版发行：大连理工大学出版社
(地址：大连市软件园路80号　邮编：116023)
印　刷：上海锦良印刷厂
幅面尺寸：225mm×300mm
印　张：13
出版时间：2018年3月第1版
印刷时间：2018年3月第1次印刷
出 版 人：金英伟
统　筹：房　磊
责任编辑：张昕焱
封面设计：王志峰
责任校对：杨　丹
书　号：978-7-5685-1385-2
定　价：258.00元

发　行：0411-84708842
传　真：0411-84701466
E-mail:12282980@qq.com
URL: http://dutp.dlut.edu.cn

本书如有印装质量问题，请与我社发行部联系更换。